# Starter Guide for
# STM32Cube™ and Eclipse ThreadX®

By
Sean D. Liming and John R. Malin

# Copyright

# Dedication

Dedicated to Gary Boone (Texas Instruments) who invented the first microcontroller, and Sophie Wilson and Steve Furber the creators of ARM.

# Table of Contents

PREFACE ..................................................................................................................... I

ACKNOWLEDGEMENTS ........................................................................................... III

ANNABOOKS ............................................................................................................. IV

1   SILICON TO CLOUD ........................................................................................... 1

   1.1   STM32 AND STM32CUBE TOOLS ............................................................... 1
   1.2   THREADX BECOMES AZURE RTOS AND THEN BECOMES ECLIPSE THREADX® ........... 2
   1.3   WHAT THIS BOOK COVERS: DRIVING QUESTIONS ............................................. 3
   1.4   SOURCE CODE LICENSE .............................................................................. 4
   1.5   CAUTION ON TOOLS AND CODE CHANGES ...................................................... 4
   1.6   PROJECTS DOWNLOAD................................................................................. 5
   1.7   SUMMARY: DIVING RIGHT IN AND FEEDBACK.................................................... 5

2   DEVELOPMENT KITS .......................................................................................... 7

   2.1   NUCLEO SERIES AND EXPANSION MODULES. .................................................. 7
   2.2   DISCOVERY SERIES ................................................................................... 7
   2.3   EVALUATION BOARDS................................................................................. 7
   2.4   ST-LINK IN-CIRCUIT DEBUGGER/PROGRAMMER ............................................. 8
   2.5   DEVELOPMENT BOARDS USED IN THE BOOK.................................................... 8
   2.6   SUMMARY: THE RIGHT DEVELOPMENT BOARD FOR YOUR PROJECT ...................... 9

3   TOOL INSTALLATION ........................................................................................ 11

   3.1   STM32CUBEMX ................................................................................... 11
   3.2   STM32CUBEIDE ................................................................................... 12
   3.3   STM32CUBE PACKAGE REPOSITORY CHECK................................................. 13
   3.4   DOWNLOAD DRIVER AND SAMPLE PROJECTS FOR TARGET BOARDS ..................... 15
   3.5   STM32CUBEPROG ................................................................................ 17
   3.6   STM32CUBEMONITOR AND MCU FINDER (OPTIONAL) ................................... 18
   3.7   SERIAL TERMINAL PROGRAM: ABCOMTERM ............................................... 18
   3.8   TRACEX .............................................................................................. 19
   3.9   TOUCHGFX DESIGNER (OPTIONAL) ............................................................ 19
   3.10   GUIX STUDIO (OPTIONAL)..................................................................... 20
   3.11   GIT .................................................................................................. 20
   3.12   DOWNLOAD GETTING STARTED EXAMPLES (GSE) ........................................ 21
   3.13   DOWNLOAD STM32CUBEIDE EXAMPLE .................................................... 22
   3.14   DOWNLOAD THREADX COMPONENTS ........................................................ 22
   3.15   OTHER TOOLS TO CONSIDER .................................................................. 23
   3.16   SUMMARY: LOW COST TO GET STARTED ................................................... 23

4   STM32 CORTEX PRIMER .................................................................................. 25

   4.1   DIFFERENT MCUS DIFFERENT ARCHITECTURES............................................. 25
   4.2   MEMORY MAP ....................................................................................... 26
   4.3   START-UP ............................................................................................ 27

4.4     CLOCKS AND TIMERS ....................................................................................... 28
4.5     DEVELOPMENT KIT DOCUMENTATION ............................................................. 29
4.6     SUMMARY: IT IS ALL IN THE DETAILS ............................................................... 29

5     PROJECT 1: BLINKING LED APPLICATION NO RTOS ......................................... 31

5.1     PART 1: INITIATE A BLINKING LED PROJECT 1 WITH STM32CUBEMX ..................... 31
5.2     PART 1: PROJECT SET UP IN STM32CUBE IDE ................................................... 34
5.3     PART 1: MODIFY THE MAIN.C TO BLINK THE LED WITH THE HAL FUNCTIONS .......... 36
5.4     PART 1: DEBUG THE APPLICATION ON THE BOARD .............................................. 40
5.5     PART 2: INITIATE THE BLINKING LED PROJECT 2 WITH STM32CUBEMX ................. 41
5.6     PART 2: PROJECT SETUP IN STM32CUBEIDE ..................................................... 44
5.7     PART 2: MODIFY MAIN.C TO BLINK AN LED USING HAL FUNCTIONS ..................... 46
5.8     PART 2: DEBUG THE APPLICATIONS ON THE BOARD ............................................ 48
5.9     SUMMARY: FIRST LOOK AT THE TOOLS ............................................................. 49

6     PROJECT 2: THREADX BLINKING LED APPLICATION ....................................... 51

6.1     INITIATE THE PROJECT WITH STM32CUBEMX .................................................... 51
6.2     THREADX FILE ADDITIONS ............................................................................. 56
6.3     EDIT THE CODE ............................................................................................. 57
6.4     DEBUG THE APPLICATIONS ON THE BOARD ...................................................... 61
6.5     ADD A SECOND THREAD ................................................................................. 62
6.6     SUMMARY: FIRST THREADX APPLICATION ....................................................... 64

7     PROJECT 3: THREADS AND TRACEX .................................................................... 65

7.1     ADD THE TRACEX SOFTWARE PACKAGE ........................................................... 65
7.2     EDIT THE CODE TO ADD THE TRACE BUFFER ..................................................... 67
7.3     DEBUG THE APPLICATION AND CAPTURE THE TRACE BUFFER .............................. 69
7.4     VIEW BUFFER DATA IN TRACEX ...................................................................... 71
7.5     SUMMARY: A VIEW INTO APPLICATION RUNNING .............................................. 73

8     PROJECT 4: BAROMETER -NO RTOS ................................................................... 75

8.1     INITIATE THE PROJECT WITH STM32CUBEMX .................................................... 75
8.2     MEMS DRIVERS VERSUS EMBEDDED FIRMWARE PACKAGE DRIVERS ................... 76
        8.2.1     Add the MEMS1 Drivers .................................................................. 76
        8.2.2     Compare Source Code ..................................................................... 79
        8.2.3     Remove the MEMS1 Drivers .............................................................. 80
        8.2.4     Add the Embedded Firmware Drivers ................................................. 80
8.3     EDIT THE CODE ............................................................................................. 82
8.4     DEBUG THE APPLICATIONS ON THE BOARD ...................................................... 84
8.5     SUMMARY: CUSTOMIZING THE PROJECT .......................................................... 85

9     PROJECT 5: NETX DUO ......................................................................................... 87

9.1     RUN THE EXAMPLE PROJECT ......................................................................... 87
        9.1.1     Import the Workspace Project ........................................................... 87
        9.1.2     Build and Debug the Ping Project ..................................................... 87
9.2     EXAMPLE AND SOFTWARE PACKAGE RESEARCH ............................................... 89
9.3     NETX DUO COMPONENT ADD-ONS ................................................................. 90
9.4     INITIATE THE PROJECT WITH STM32CUBEMX .................................................... 92

9.5 ADD NETX DUO FROM STM32CUBEIDE EXAMPLES.................................97
9.6 ADD THE WI-FI DRIVER...........................................................98
9.7 SET THE INCLUDE PATHS........................................................99
9.8 EDIT THE CODE....................................................................101
    *9.8.1 Create a Wi Fi Setup File ...............................................101*
    *9.8.2 Initialize NetX Duo ....................................................107*
    *9.8.3 Add Call to initialize Wi-Fi. .........................................109*
    *9.8.4 Define printf() and scanf() calls to go out and in the debug port ......110*
    *9.8.5 Add Random Number Generator Code ................................111*
9.9 BUILD AND DEBUG................................................................112
9.10 SUMMARY: ONE STEP CLOSER TO THE CLOUD ...........................113

**10 PROJECT 6: AZURE IOT CENTRAL CONNECTION ..........................................115**

10.1 REVIEW: THE TWO-SAMPLE AZURE RTOS PROJECTS. ....................115
10.2 INITIATE THE PROJECT WITH STM32CUBEMX ...........................116
10.3 ADD SENSOR AND WIFI DRIVER SOURCE CODE............................119
10.4 ADD MISSING NETX DUO ADDONS AND SEPARATE THE SAMPLE.............123
10.5 EDIT THE CODE FOR MAIN.C ................................................126
10.6 ADD DEFINES TO NX_PORT.H ...............................................129
10.7 ADD THE AZURE IOT SAMPLE SOURCE CODE FROM THE NETX DUO COMPONENT
DOWNLOAD ...........................................................................130
10.8 CREATE AZURE IOT CENTRAL APPLICATION ...............................135
10.9 BUILD AND DEBUG...............................................................137
10.10 SUMMARY: QUESTION ANSWERED ..........................................144

**11 PROJECT 7: FILEX ........................................................................145**

11.1 USING RAM, CREATE PROJECT PART 1 WITH STM32CUBEMX ..............145
11.2 FILEX ADDITIONS TO THE PROJECT .........................................151
11.3 WRITING THE APPLICATION. ..................................................152
11.4 DEBUG THE APPLICATIONS ON THE BOARD ................................156
11.5 USING NOR FLASH, CREATE PROJECT PART 2 WITH STM32CUBEMX.........158
    *11.5.1 Why not use the STM32L4S5 Discovery Kit? .........................158*
    *11.5.2 Moving forward with STM32U5 Discovery Kit .......................159*
11.6 FILEX AND LEVELX ADDITIONS TO THE PROJECT ..........................165
11.7 WRITING THE APPLICATION. ..................................................166
11.8 DEBUG THE APPLICATIONS ON THE BOARD ................................169
11.9 SUMMARY: STORAGE – DETAILS IN THE SETUP .............................170

**12 PROJECT 8: NUCLEO-H723 NETX DUO ...............................................171**

12.1 CREATE A PROJECT WITH STM32CUBEMX ................................171
12.2 WRITING THE APPLICATION. ..................................................179
12.3 MODIFYING THE LINKER SCRIPT .............................................181
12.4 DEBUG THE APPLICATIONS ON THE BOARD ................................183
12.5 SUMMARY: ALL THE SMALL THINGS .........................................184

**13 PROJECT 9: DUAL-CORE ..............................................................185**

13.1 CREATE DUAL-CORE WITH STM32CUBEMX ................................185
13.2 MODIFY THE MAIN.C FILES ....................................................189

13.3    SETTING UP THE DEBUG CONFIGURATION FOR BOTH PROJECTS..................193
13.4    RUNNING THE DEBUGGER..................197
13.5    ADD AZURE RTOS TO BOTH CORES..................198
    13.5.1    Project Changes..................198
    13.5.2    Modify the Source Code..................201
    13.5.3    Debug the Code..................202
13.6    SHARED MEMORY EXAMPLE..................203
13.7    SUMMARY: TWO CORES BETTER THAN ONE..................207

14    PROJECT 10: THREADX AND OPENAMP..................209

14.1    CREATE THE OPENAMP PROJECT WITH STM32CUBEMX..................209
14.2    MODIFY THE SOURCE CODE..................218
    14.2.1    Apply a Patch..................218
    14.2.2    CM4 Subproject..................219
    14.2.3    CM7 Subproject..................221
14.3    DEBUG THE CODE..................224
    14.3.1    Debug Set Up..................224
    14.3.2    Running the Debugger..................226
14.4    SEND DATA FROM SLAVE TO MASTER..................228
14.5    SUMMARY: DUAL-CORES WORKING TOGETHER..................229

15    GRAPHICAL USER INTERFACE INTRODUCTION..................231

15.1    WHERE IS GUIX IN STM32CUBE AND WHAT IS TOUCHGFX?..................231
    15.1.1    GUIX and TouchGFX Similarities..................231
    15.1.2    GUIX..................232
    15.1.3    TouchGFX..................232
15.2    PROJECT 11 CREATE PROJECT IN TOUCHGFX DESIGNER..................233
    15.2.1    Create the IDE in TouchGFX Desinger..................233
    15.2.2    Edit the Code..................241
    15.2.3    Run the Code with the Debugger..................246
15.3    PROJECT 12 CREATE A GUIX APPLICATION..................248
    15.3.1    Final GUIX Setup Steps..................248
    15.3.2    Create a GUI using GUIX Studio..................251
    15.3.3    Building the Libraries..................258
    15.3.4    Creating a Win32 Application..................260
15.4    GUIX INTEGRATION INTO STM32CUBEIDE PROJECT..................268
    15.4.1    Rebuild the GUIX Application for the Processor..................269
    15.4.2    Additional Changes to the Project..................271
    15.4.3    Interesting LCD Testing Results..................272
15.5    SUMMARY: THE TALE OF TWO GUI APPROACHES..................273

16    MXCHIP® IOT DEV KIT..................275

16.1    MXCHIP IOT DEV KIT OVERVIEW..................275
16.2    GETTING STARTED EXAMPLE FOR MXCHIP IOT DEV KIT REVIEW..................276
16.3    PROJECT 13 SENSORS, LEDS, BUTTONS, AND DISPLAY: CREATE THE PROJECT
WITH STM32CUBEMX..................277
    16.3.1    Enabling the I/O..................278
    16.3.2    Enable Azure RTOS ThreadX Software Package..................285

*16.3.3 Name the Project and Generate the Code* .................................................. *286*
*16.3.4 Add the Drivers to the Project* ..................................................................... *287*
*16.3.5 Edit the Driver Code* .................................................................................... *289*
*16.3.6 Final code edits* ............................................................................................. *290*
*16.3.7 Build and Debug with ST-LINK OpenOCD* ............................................. *296*
16.4 PROJECT 14: CLOCK CONFIGURATION CHANGES ............................................ 298
16.5 PROJECT 15: ADD WiFi AND NetX DUO .......................................................... 300
*16.5.1 Enable Azure RTOS NetX Duo Software Package* ................................... *300*
*16.5.2 Add the Driver to the Project* ...................................................................... *301*
*16.5.3 Add Source Files from the Getting Started Example* .............................. *303*
*16.5.4 Edit the code* .................................................................................................. *303*
*16.5.5 Build and Debug the WiFi Additions* ......................................................... *304*
*16.5.6 WiFi Investigation* ......................................................................................... *305*
16.6 PROJECT 16: BACK OUT WiFi CHANGES AND MODIFY CODE A LITTLE MORE ...... 305
*16.6.1 Back out the WiFi Changes* ......................................................................... *306*
*16.6.2 Modify the Code* ........................................................................................... *306*
*16.6.3 Build and Debug the Updated Project* ...................................................... *313*
16.7 SUMMARY .......................................................................................................... 313

17 FINAL REVIEW ...................................................................................................... 315

17.1 SUPPORT AND THE ST COMMUNITY ................................................................ 315
17.2 SUMMARY OF STM32CUBE AND STM32 DEVELOPMENT HARDWARE ............... 315
17.3 SUMMARY OF ThreadX (AZURE RTOS) AND AZURE IoT C SDK ..................... 316
17.4 FINAL SUMMARY ................................................................................................ 317

A REFERENCES ......................................................................................................... 319

B INDEX ..................................................................................................................... 323

# Preface

In my article, "Azure RTOS and STMicroelectronics STM32 Discovery Kit IoT (STM32L4S5)", I originally planned to add a final section covering how to build the Azure RTOS Getting Started Example (GSE) using the STM32CubeIDE. The more I studied the GSE and the STM32Cube tool family, it became very clear that more research and more written words were needed to explain how to make the GSE and tools work together. There was also this nagging question that STMicroelectronics has many STM32 MCU products, and they should all be able to support running Azure RTOS. After many months and exhaustive research, what was to be a simple final section to a little article turned into this book.

Although Computer Engineering was my focus in college, my profession has focused on software. More specifically, I have focused on integrating Microsoft operating systems like DOS and Windows into Embedded/IoT for over a quarter century. During COVID lockdown, I had some time to look at other areas that piqued my interest. MCUs and FPGAs have been an interest of mine since they go hand-in-hand with my college degree. There are many little devices that can be built that a PC platform just cannot do. Yes, I did support Windows CE when it first came out, but the focus was Windows CE on the PC platform. Also, I could see the writing on the wall that Windows CE was eventually going to be phased out a decade before it did.

With Windows CE end-of-life, all the Windows CE supporters have gone on to newer adventures. There are very few Microsoft MVPs left who can review or comment on the new direction from Microsoft. Since I appear to be the last one standing, here I go into a new arena. Similar to my other books, the approach to this book is to assist readers that are new to the subject. Being new myself, allowed me to catch all the little trappings that an expert might gloss over, which I found to be an issue with all the research. I want to make it clear I am not an MCU or ThreadX (Azure RTOS) expert. I am sharing what I learned on my little journey into this world.

As the book was just about to go through the edit cycle, Microsoft made the decision to pass Azure RTOS to the Eclipse Foundation for continued support. The RTOS name changed back to ThreadX. In addition, Azure IoT Central will be deprecated in a few years. There is a strong push to move away from Azure IoT Hub and Azure IoT Central and move to Azure Event Grid. Connecting to Azure IoT Central is still available as of this writing.

The key words in the book title are "Starter Guide...". The book is intended to help those getting started with ThreadX and STM32MCUs.   If you are an expert on STM32 programming, this book is not for you, but I am more than happy to receive any feedback on what I got right and wrong in my approach and information. I don't consider this book 100% complete. There is room to make a second edition so I would like to hear anyone's thoughts. Please reach out to me via Annabooks.com.


Sean D. Liming
Rancho Mirage, CA

# Acknowledgements

Playing in a new arena means working with new people that I have never interacted with before. The STMicroelectronics Community site provided the biggest assistance. Since everyone has a user profile name, it is difficult to acknowledge anyone specifically. A general thanks to the ST Community for all the responses to my questions.

# Annabooks

Annabooks provides a unique approach to embedded/IoT system services with multiple support levels. Our different offerings include books, articles, training, and project consulting. Current publications and courses have focused on embedded PC architecture and Windows Embedded, which reach a wide audience from Fortune 500 companies to small organizations. We will continue to expand our future services into new technologies and unique topics as they become available

**Books and eBooks**

Starter Guide to Windows 10 IoT Enterprise

Java and Eclipse for Computer Science

Open Software Stack for the Intel® Atom™ Processor

Professional's Guide to POS for Windows Runtime

Professional's Guide to POS for .NET

Real-Time Development from Theory to Practice

The PC Handbook

**Training Courses**

Windows® 10 IoT Enterprise Training Course

Please contact us for more information on consultation and availability.

Web: www.annabooks.com

# 1   Silicon to Cloud

Whatever you want to call it, the Internet of Things, cloud computing, or connected embedded systems, the cloud has exploded with possibilities. Devices and sensors can talk to each other in a factory over a local network and send data around the world over the Internet. Production management, device servicing, environmental controls, and system management have all been enhanced with internet connectivity. Smart phones, desktop computers, and tablets were the first devices to take advantage of what the cloud offers, but the marketing analysis shows that the number of smaller devices connecting to the cloud will be much greater than their general computing counterparts. Silicon vendors and Independent Software Vendors (ISVs) are coming up with solutions to help developers deliver cloud-based solutions in shorter development times. This book explores two vendor solutions: STMicroelectronics STM32Cube™ tools and Eclipse ThreadX®.

## 1.1   STM32 and STM32Cube Tools

Founded in 1987, STMicroelectronics has become a global silicon device manufacturer. ARM processors grew rapidly in the late 1990s and into the 2000s. Starting in 2007, STMicroelectronics created the STM32 MCU product line based on the 32-bit ARM Cortex-M core. Different family-series of MCUs are based on different ARM Cortex-M cores.

| STM32 MCU Family | ARM Cortex-M Core(s) |
|---|---|
| F0 | Cortex-M0 |
| C0, G0, L0 | Cortex-M0+ |
| F1, F2, L1 | Cortex-M3 |
| F3, F4, G4, L4, L4+ | Cortex-M4 |
| WB, WL | Cortex-M4 & Cortex-M0 (2 cores) |
| F7 | Cortex-M7 |
| H7 | Cortex-M7 (1 Core), Cortex-M7 & Cortex-M4 (2 Cores) |
| H5, L5, U5, WBA | Cortex-M33 |

MCUs have multiplexed pins and integrated clock signaling. Configuring the registers and writing the code for each I/O type can be time-consuming. Third-party tool vendors initially provided firmware development support; but in 2014, STMicroelectronics launched the STM32Cube tools. This family of tools helps with the initial development of devices through programming and debugging.

- STM32CubeMX – MX is a graphical tool to help configure all the available STM32 MCUs. You can define each pin, set up the different clocks, and add software packages. Once the selection is completed, code is generated, which takes care of all the low-level configuration details for the project.

- **STM32CubeIDE** – The IDE, Integrated Development Environment, is the main project development tool and is based on Eclipse. This is the main tool for project development and debugging.

- **STM32CubeProgrammer** – All STM32 MCUs have a built-in boot loader that is used to download code to the internal flash via a UART port. The Programmer (Pro) supports programming through the USB-UART. The Programmer can also read, write, and verify MCU memory via JTAG/SWD and Bootloader.

- **STM32CubeMonitor** – A debug tool that provides a solution to monitor and diagnose applications at runtime.

The tools provide the Hardware Abstraction Layer (HAL) source code, startup code, and memory linker file so you can focus on the application. As with many MCUs, the STM32 has support from different real-time operating systems such as FreeRTOS, emdOS, OpenSTLinux, and ThreadX; so, you can create multithreaded applications. When Microsoft acquired Express Logic, the makers of ThreadX, STMicroelectronics made an investment to have integrated support for ThreadX.

## 1.2 ThreadX becomes Azure RTOS and then becomes Eclipse ThreadX®

In 1997, Express Logic launched ThreadX. Developed by William Lamie, ThreadX's small footprint and extended capability quickly became a widely used RTOS for a number of applications. Like other RTOSes, ThreadX is a preemptive multitasking operating system with interrupt response, memory management, and interthread communications. ThreadX has achieved a number of certifications to address safety critical systems. With a market projection of billions of MCU devices connected to the cloud, major cloud providers had to make some strategic decisions. Amazon was the first to jump in and acquired FreeRTOS. Amazon created a new version called AWS FreeRTOS that can connect to AWS cloud services. Since Windows CE went end-of-life with the 2013 release, Microsoft had to fill the gap to support something for MCUs. In 2019, Microsoft acquired Express Logic and renamed ThreadX to Azure RTOS. Besides providing a few examples of connecting to Azure and making Azure RTOS license free, communication on where Microsoft was going to go with Azure RTOS was limited. As it turns out the engineering support needed to maintain Azure RTOS was not something that fit the standard Microsoft software resources. In 2024, Azure RTOS was moved to the Eclipse Foundation for the open-source community to support. The product is now called Eclipse ThreadX® or simply ThreadX®.

The change happened as we started the editing cycle for the book. All the projects and screen captures show Azure RTOS. Since the STM32Cube tools cannot be changed overnight, you will see a mixed of the names (ThreadX, Azure RTOS, ThreadX(Azure RTOS)) throughout the book. We have attempted to make this as clear as possible, but this confusion was out of our control.

2

Besides all the changes, Eclipse ThreadX is focused on the latest MCUs from different silicon vendors, Azure connectivity support as expected, and a different licensing model. The six components that make up ThreadX are the same:

- ThreadX – This is the real-time kernel that all other components require to run.

- FileX – Provides FAT file system support with fault tolerance protections and wear-level support for flash devices.

- USBX – USB stack to act as both a client and host device.

- NetX and NetX Duo – Provides a TCP/IP stack along with a number of add-on network and security technologies such as TLS, MQTT, PPP, SNTP, DCHP, Webserver, and much more.

- GUIX and GUIX Studio – A graphical design and runtime environment for 2D graphic applications.

- TraceX – A graphical analysis tool for viewing tasks and interrupt switching to help solve system-level behavior problems and performance tuning.

There is an unofficial 7th component called the Azure IoT C SDK that has the code to connect to Azure IoT Hub and DPS services. In reality, the 7th component is just an add-on to NetX Duo.

## 1.3 What this Book Covers: Driving Questions

After the acquisition and rebranding to Azure RTOS, Microsoft also released two different types of example projects that demonstrate creating an image that runs on a board and connects to Azure IoT Central or Azure IoT Hub. The first is the Azure RTOS Getting Started Example (GSE), which is a complete solution to send sensor data to Azure IoT Central. The whole project is built with Visual Studio Code, which requires several packages and 3rd party components to be installed. The second example is composed of individual sample projects that work with 3rd party and silicon vendor development tools like STM32Cube. The second example projects create an image that sends dummy sensor data to Azure IoT Hub. Both example projects support a variety of development boards from silicon vendors. For STMicroelectronics, three different MCU boards are used in these two project sets. These examples are fine to get familiar with ThreadX projects connecting to Azure IoT, but there are many different MCUs in the STM32 product line. The main driving questions for creating this book are:

1. **How do you get started from scratch to create a solution that connects to Azure IoT Hub or Azure IoT Central that matches the current examples from Microsoft?**

2. **How do you create a ThreadX (Azure RTOS) project for different STM32 MCUs that are not supported by any of the example sets?**

After reviewing the different silicon vendor tools, it was clear that STMicroelectronics made investments to deliver Azure RTOS for their silicon making it easy to choose STM32 MCUs to help answer these driving questions. As the title of the book implies, this is a starter guide. The book does NOT cover RTOS theory, the RTOS features of ThreadX, or writing device drivers. The book's focus is on how to create a ThreadX (Azure RTOS) project from scratch using the STM32Cube tool suite and how to connect to Azure IoT using the already available source code. The book covers 5 of the 6 different ThreadX components. Along the way, different features of STM32Cube tools will be covered.

The book is designed for hands-on learning. All the projects focus on step-by-step examples. Chapters 2 and 3 cover the STM32 development boards and the tools needed for development. Chapter 4 provides a high-level overview of the STM32 MCU Cortex architecture. Chapters 5 through 9 are different projects that walk through the basic elements of the STM32Cube tools and implementing ThreadX. The culmination of the previous chapters is chapter 10, which is the complete project that connects to Azure IoT Central. Chapter 11 provides examples for FileX. Chapter 12 demonstrates networking over a physical ethernet port. Chapters 13 and 14 walk through the steps to set up Azure RTOS for dual-core STM32 MCUs. Chapter 15 introduces GUI programming. Chapter 16 covers a reference board from a third-party vendor and introduces some concepts that are covered in earlier chapters.

## 1.4   Source Code License

The different example projects and code generated by STM32Cube tools have different licenses called out in the code. Some have a Microsoft license, an MIT license, and an STMicroelectronics license. Currently, the Eclipse ThreadX project is under the MIT license. Different code snippets will be re-used for learning purposes. If you intend to reuse any of the supplied code from the production examples, you must include the relevant license information in the header of the source code.

Any code that is not covered by the license agreements is free to use but is supplied as-is, without warranty.

## 1.5   Caution on Tools and Code Changes

As time goes on, the example project and tools will change over time. The book is based on what was available at the time of writing. The STM32Cube tools had several updates during the book's development, but the changes appeared to be minor or cosmetic. Please be aware that some of the pictures and code examples might not align with the present state of the tools.

## 1.6   Projects Download

A Zip file containing the different projects in the book can be download from the book's page at Annabooks.com. Per the previous section, the STM32Cube tools will change so the project may have to be upgraded to the latest release.

## 1.7   Summary: Diving Right in and Feedback

Each project chapter jumps right into using the tools to develop the project. Different features and nuances of the tools and ThreadX are covered along the way. It is impossible to cover everything, so your feedback on what you think is missing, important, or can be improved for future editions or a more advanced text is appreciated.

# 2  Development Kits

Any good technology company provides software demos or example hardware so developers get familiar with that company's products. STMicroelectronics does a very good job in providing development boards for many of their MCUs. Different MCUs come in different packages which change the number of pins, but the development board offering cuts across all of the STM32 MCU family. The goal is to allow application development to make progress with the development board while real custom hardware is developed. STMicroelectronics offers different series of development boards to address one's development needs.

## 2.1  NUCLEO Series and Expansion Modules.

The NUCLEO series provides the most flexible prototyping solution. NUCLEO development boards come in 3 flavors based on pin package count: NUCLEO-32, NUCLEO-64, and NUCLEO-144. The boards provide an ST-Link connection for development and the MCU with all the pins exposed to header connections. Some NUCLEO-144 boards have a physical Ethernet connection. There are a number of expansion modules that plug into the header connectors providing sensors, communication, motor control, displays, and multimedia.

## 2.2  Discovery Series

The discovery series provides a more complete solution with different industry expansion options such as Arduino Uno, STmod+, MikroBus, and Grove. Many of the development boards in this series include sensors, displays, audio, USB, and wireless connectivity on the board.

## 2.3  Evaluation Boards

The evaluation boards are for the high-performance STM32 MCUs. The boards provide the basic connections for RS232, I2C, and SPI, as well as, some sensors, USB connections, and HDMI / LCD interfaces. There is a daughterboard expansion connection available for custom boards. Debugging support for ST-Link, as well as, JTAG/SWD is also available.

## 2.4   ST-Link In-Circuit Debugger/Programmer

Most of the development boards include an ST-Link interface for in-circuit debugging on board. ST-Link communicates with the internal boot loader of the MCU to program the flash and provides a software debugging interface. ST-Link is simply another STM32 MCU that has been programmed to interface with the target MCU. For boards without the built ST-Link, STMicroelectronics offers inexpensive ST-Link in-Circuit debugger programmers that can connect to the SWD/JTAG port. When you develop a custom board, you will have to expose the serial-wire-debug (SWD)/JTAG interface for an external ST-Link in-circuit debugger.

## 2.5   Development Boards Used in the Book

The different projects in the book will use different STM32 MCU development boards as the target hardware. By using different development boards, the goal is to demonstrate different features of STM32Cube tools, ThreadX components, and answer the book's driving questions. Here is the list of the development platforms used in the book:

- The B-L4S5I-IOT01A Discovery kit provides a development platform for the STM32L4S5VI core microcontroller.
- The B-U585I-IOT02A Discovery kit provides a development platform for the STM32U585AI core microcontroller
- The NUCLEO-H723ZG is the STM32 Nucleo-144 development board with STM32H723ZG MCU.
- The STM32H747I-DISCO Discovery kit provides a development platform for the STM32H747XIH6 microcontroller complete with 4" capacitive touch LCD display module with MIPI® DSI interface.
- MXCHIP IoT DevKit

The first 4 boards can be purchased from the STMicroelectronics website. The MXCHIP IoT DevKit is a little harder to find. Here is the summary of each of the platforms:

The STM32L4S5-based Discovery Kit BL-4S5I-IOT01A and STM32U585AI-based Discovery Kit B-U585I-IOT02A were used to develop the ThreadX (Azure RTOS) project examples from Microsoft. Since we wanted to repeat the examples from scratch, it makes sense to have working projects to compare against. The STM32L4S5 is the primary board for chapters 4-9, since it is older and offers different learning options when it comes to the tools. The STM32U585AI-based Discovery Kit B-U585I-IOT02A is newer and is closer to working out-of-the-box with the STM32Cube tools. The STM32U585AI-based Discovery Kit B-U585I-IOT02A is going to be used in the second half of the FileX project.

The NUCLEO-H723ZG was used in the STMicroelectronics online seminar that is found on YouTube. This basic board is used to demonstrate FileX and wired networking.

The STM32H747XIH6-based STM32H747I-DISCO Discovery kit uses an ARM dual-core processor. The original reason for choosing this platform was that there was an example GUIX application developed by someone in the ST Community. After studying the board, it became evident that there was more to learn and share. First, setting up the debugging for this board is a little challenging, but a good learning exercise on how to debug dual-core systems. Second, how two cores interact with each other to share data is an important concept of a dual-core system. Finally, the kit also comes with an LCD touch display screen which can be used for 2D graphical applications. We will cover two GUI library solutions that can be used with this display screen.

MXCHIP IoT DevKit is one of the original IoT DevKits that Microsoft used to demonstrate connecting to Azure IoT. The board uses an STM32F412 MCU, but the implementation is a little different. The board is a little older, so you might not be able to find it, but Chapter 16 is used to demonstrate creating a project from the MCU and not the reference board.

## 2.6   Summary: The Right Development Board for Your Project

With so many MCU, development board, and development kit options available, only you can pick the right combination that fits your application. STMicroelectronics has created a number of development boards and development kits that allow you to get to prototyping and up the learning curve quickly. The development boards and development kits chosen for the book have available projects to fall back on. You are more than welcome to use a different STM32 development board. Keep in mind the steps and code might be different for a different STM32 development board.

# 3 Tool Installation

There are several software development tools that will be needed and some that are optional for building the different projects. The most important tools are from STMicroelectronics. The STM32 tool chain is comprised of several software tools that help with getting started, code development, board programming, and optimizations. This chapter covers the installation of all the tools.

## 3.1 STM32CubeMX

The STM32CubeMX is the getting started tool for development. STM32CubeMX provides a wizard to walk through the steps to create a project and configure the I/O for the MCU or board you have selected. You can start any project with the STM32CubeIDE and walk through the same wizard to create the project, but the STM32CubeMX has support for IAR Embedded Workbench and Keil MDK if you prefer those tool chains. For this book, we will focus on the STM32CubeIDE.

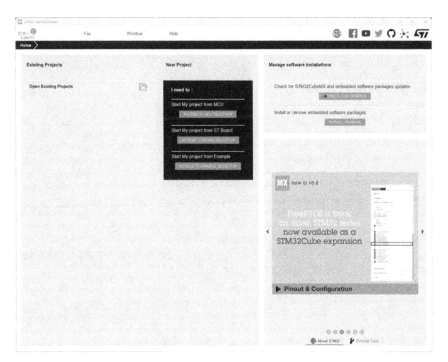

1. Download STM32CubeMX from STMicroelectronics: https://www.st.com/en/development-tools/stm32cubemx.html.
2. Extract the contents from the ZIP file.
3. Run the installer to install STM32CubeMX.
4. Choose the defaults to finish the installation.

## 3.2   STM32CubeIDE

The STM32CubeIDE is the main development tool to compile the project into a binary that can be downloaded to the board. The STM32CubeIDE is built on the Eclipse®/CDT™ so familiarity with Eclipse development comes in handy.

1. Download the STM32 Cube IDE from STMicroelectronics: https://www.st.com/content/st_com/en/products/development-tools/software-development-tools/stm32-software-development-tools/stm32-ides/stm32cubeide.html .
2. Extract the contents from the ZIP file.
3. Run the installer to install STM32CubeIDE.
4. Choose the defaults to finish the installation.
5. Once the installation has been completed, run STM32CubeIDE.
6. First, you will be asked to choose a workspace folder for your projects. It is recommended to put the workspace folder in a simple easy-to-access location.

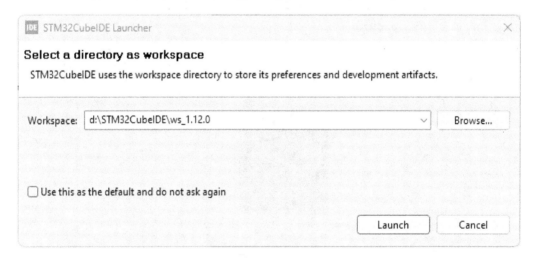

7. Click the Launch button when finished.
8. You may be presented with further configuration questions. As you can see, there are options to start a project from the IDE.

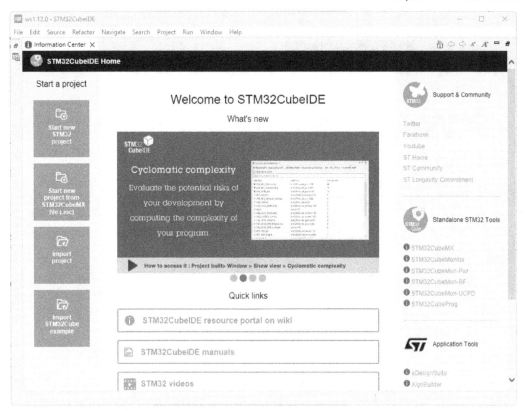

## 3.3 STM32Cube Package Repository Check

The STM32Cube tools come with a source code software package to help get a design up and running faster. The packages are stored in c:\Users\<user account>\STM32Cube\Repository. Both STM32CubeMX and STM32CubeIDE will share this repository. To make sure that both tools are pointing to the same path:

1. Open STM32CubeMX.
2. From the menu select Help->Updater Settings.
3. A dialog appears showing the path to the repository:

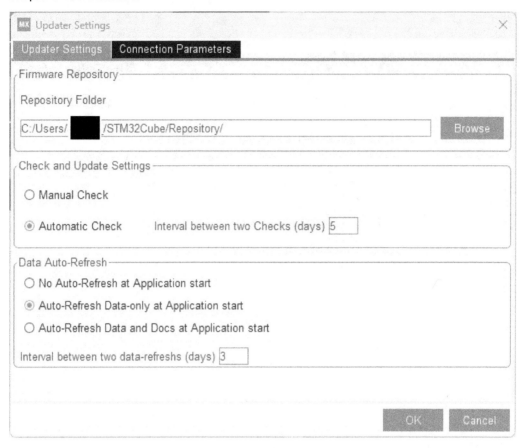

4. Open STM32CubeIDE
5. From the menu select Windows->Preferences
6. A dialog appears, select STM32Cube->Firmware Updater. The Repository path should match:

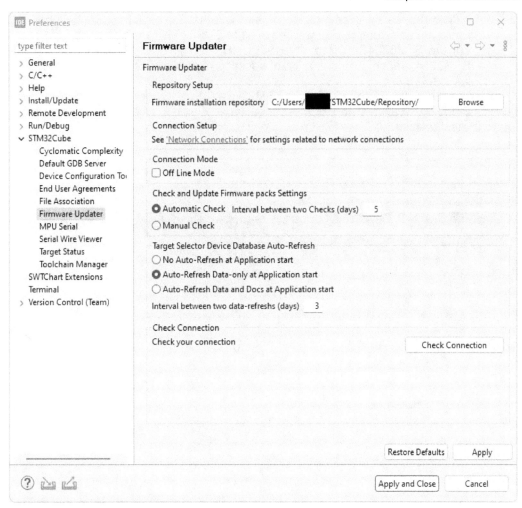

5. Close both dialogs and Close STM32CubeIDE.

## 3.4 Download Driver and Sample Projects for Target Boards

The source code software packages that are available are only part of the story. There are other embedded firmware packages that contain other device drivers and example projects that can be used for learning or as a starting point for a custom design. Not to get too far ahead of the story, yes, two different source code packages imply different solution implementations. For example, the MEMs Driver package for different sensor components implements the driver differently than the firmware packages. We will show this in a later chapter. Let's download the two embedded firmware packages for the selected development kits.

1. Open STM32CubeMX.
2. From the menu, select Help -> "Manage embedded software packages".

3. A dialog appears with several tabs. Make sure that you have the STM32Cube MCU Package tab selected. You will see a list of STM32 MCU names by family.
4. For the STM32L4S5 Discovery Kit (BL-4S5I-IOT01A), scroll to and expand the STM32L4 branch. There is a list of package releases. Check the box next to the latest release.
5. For the NUCLEO-H723ZG and STM32H747I-DISCO, scroll to and expand the STM32H7 branch. There is a list of package releases. Check the box next to the latest release.

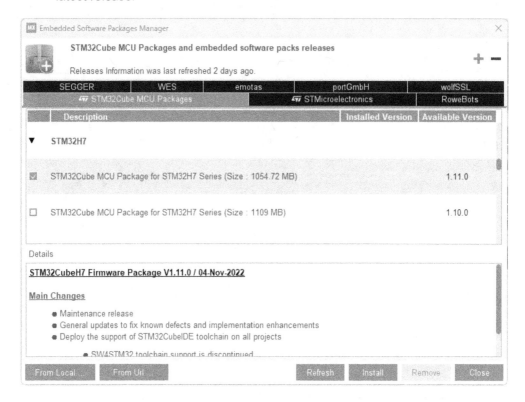

6. Click Install. The packages will be downloaded and unzipped to the repository.
7. A dialog will appear asking for you to accept the license agreement. Check the accept button and click Finish. The package downloading and unzipping will be completed.
8. Close the dialog when finished.
9. Close STM32CubeMX.
10. Open File Explorer.
11. Navigate to the c:\Users\<user account>\STM32Cube\Repository folder.
12. You will see the two package folders:

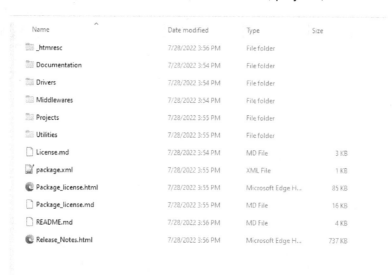

| Name | Date modified | Type | Size |
|---|---|---|---|
| Packs | 7/28/2022 3:58 PM | File folder | |
| STM32Cube_FW_H7_V1.11.0 | 4/25/2023 4:30 PM | File folder | |
| STM32Cube_FW_L4_V1.17.2 | 7/28/2022 3:55 PM | File folder | |
| ad.zip | 4/12/2023 6:33 PM | Compressed (zipp... | 550 KB |
| ad01.png | 4/12/2023 6:33 PM | PNG File | 106 KB |
| ad02.png | 4/12/2023 6:33 PM | PNG File | 117 KB |
| ad03.png | 4/12/2023 6:33 PM | PNG File | 124 KB |

13. Open one of the package folders and you will see a directory with sample documentation, driver source code, middlewares, projects, etc.

| Name | Date modified | Type | Size |
|---|---|---|---|
| _htmresc | 7/28/2022 3:56 PM | File folder | |
| Documentation | 7/28/2022 3:54 PM | File folder | |
| Drivers | 7/28/2022 3:54 PM | File folder | |
| Middlewares | 7/28/2022 3:54 PM | File folder | |
| Projects | 7/28/2022 3:55 PM | File folder | |
| Utilities | 7/28/2022 3:55 PM | File folder | |
| License.md | 7/28/2022 3:54 PM | MD File | 3 KB |
| package.xml | 7/28/2022 3:55 PM | XML File | 1 KB |
| Package_license.html | 7/28/2022 3:55 PM | Microsoft Edge H... | 85 KB |
| Package_license.md | 7/28/2022 3:55 PM | MD File | 16 KB |
| README.md | 7/28/2022 3:56 PM | MD File | 4 KB |
| Release_Notes.html | 7/28/2022 3:56 PM | Microsoft Edge H... | 737 KB |

The documentation walks through the contents of the package. It is good to walk through the different projects to get a different approach to developing a solution. We will be referring to these packages in later chapters.

14. Close Explorer when finished.

## 3.5 STM32CubeProg

The STM32CubeProg is both a graphical interface and a Command-line interface for programming STM32 products. The tool allows you to access memory and to program different internal memory devices.

1. Download the STM32 Cube Programmer from STMicroelectronics: https://www.st.com/en/development-tools/stm32cubeprog.html.
2. Extract the contents from the ZIP file.
3. Run the installer to install STM32CubeProg.
4. Choose the defaults to finish the installation. There will be a separate driver installer for STLink. Just choose the defaults to install the driver.

## 3.6   STM32CubeMonitor and MCU Finder (Optional)

There are some additional development tools in the STM32 software development family that we will not be covering in the book. The first is the STM32CubeMonitor. It is a remote monitoring program to help fine-tune your application. The MCU finder helps you pick the right STMicroelectronics microprocessor for your platform based on your custom search criteria.

## 3.7   Serial Terminal Program: ABCOMTERM

A serial terminal program is needed to receive output messages from the development board. You are more than welcome to use Putty or HyperTerminal. Annabooks has created a no-nonsense terminal program called ABCOMTERM, which can also be used.

1. Download ABCOMTERM from Annabooks.com (Annabooks COM Terminal) or download the serial terminal program of your choice.
2. Install the serial terminal program.

## 3.8 TraceX

TraceX is a post-mortem analysis tool designed to view thread sequences that run during the execution of the application.

1. Open the Microsoft Store App.
2. Search on TraceX.
3. Open the page to TraceX and click the Install button.

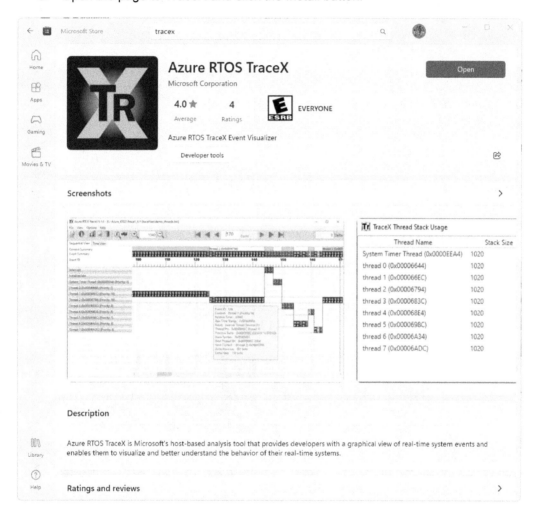

## 3.9 TouchGFX Designer (Optional)

STMicroelectronics has developed a GUI framework API to support GUI applications on STM32 MCUs. TouchGFX Designer is used to develop the GUI application and an STM32CubeIDE project for a development board.

1. Open a web browser.
2. Go to the following website: https://www.st.com/en/development-tools/touchgfxdesigner.html or search for TouchGFX download.
3. Download the latest version.
4. Extract the ZIP file.
5. In the location of the unzipped package, go to Utilities\PC_Software\TouchGFXDesigner and run the TouchGFX-4.x.x.msi file to install TouchGFX Designer.

## 3.10 GUIX Studio (Optional)

ThreadX also has GUIX tool to create graphical user interfaces. There will be a brief introduction in a later chapter. The GUIX Studio is used to create the interface and generate C files that can be integrated into either a PC project or an MCU project.

1. Open the Microsoft Store application.
2. Search for "GUIX".
3. Select Azure RTOS GUIX Studio from the results.
4. Click the install button to install the application.

The application will set up a workspace under C:\Azure_RTOS\GUIX-Studio-6.1. All projects will be stored under that folder. Visual Studio 2022 will also be needed for the GUIX project presented in Chapter 15.

## 3.11 Git

Various source code samples will be downloaded from GitHub.

**Note**: The utility gets updated often, so the following steps might change over time:

1. Download the Git utility from https://git-scm.com/downloads .

2. Run the installer
   a. Accept the license and click Next.
   b. Leave the install location as is and click Next.
   c. Leave the Selected Components as they are and click Next.
   d. Keep the State Menu Folder as is and click Next.
   e. Set the default editor to be the editor of your choice and click Next.

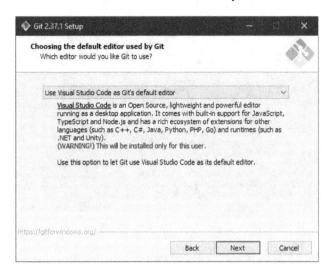

   f. Keep the default for initial branches and click Next.
   g. Keep the default PATH Environment and click Next.
   h. Keep the default OpenSSH selection and click Next.
   i. Keep the default OpenSSL selection and click Next.
   j. Select "Use Windows' default console window" and click Next.
   k. Keep the defaults for the next question and click Next.
   l. Select "Use Windows' default console window" and click Next.
   m. Keep the defaults for the next question and click Next.
   n. Keep the defaults for the next question and click Next.
   o. Keep the defaults for the extra options and click Next.
   p. Keep the defaults for the experimental options and click Install.
   q. Click Finish once the installation is completed.

## 3.12 Download Getting Started Examples (GSE)

With Git installed, we can download the getting started examples from Eclipse. The download contains support for multiple MCU manufacturers. It includes a working example to send data to Azure IoT Central. We will use this later to get code for future projects.

1. Create a directory called \Azure-RTOS-STM32
2. Open PowerShell.
3. Change directory to the newly created folder:

```
cd  \Azure-RTOS-SM32
```

4. Run the following

```
git clone --recursive https://github.com/eclipse-threadx/getting-
started
```

## 3.13  Download STM32CubeIDE Example

There is a second set of examples that we will draw solutions from.

1. Open a web browser.
2. Go to the following page: https://github.com/eclipse-threadx/samples
3. In the table that lists different IDE examples, click on the STM32CubeIDE link for the B-L4S5I-IOT01A.
4. Put the download in the \Azure-RTOS-STM32 directory.
5. Extract the zip file.

## 3.14  Download ThreadX Components

We will also want to download some of the ThreadX components from GitHub.  Some projects draw from or use the components from GitHub.

1. Create a subdirectory under \Azure-RTOS-STM32 called Components.
2. Open Power Shell.
3. Change directory to the Azure-RTOS-STM32\Components.
4. Run the following commands:

git clone --recursive https://github.com/eclipse-threadx/threadx

git clone --recursive https://github.com/eclipse-threadx/netxduo

git clone -recursive https://github.com/eclipse-threadx/filex

We will not be covering USBX in this book. GUIX will be introduced and downloaded in Chapter 15.

## 3.15 Other Tools to Consider

A few more tools to consider are Notepad++, WinMerge, and Visual Studio 2022.

- Notepad++ - https://notepad-plus-plus.org/
- WinMerge - https://winmerge.org/ - WinMerge is helpful to compare code listings.
- Visual Studio 2022 will be needed for the GUIX chapter.

Finally, if you are building your own board, you might want to consider a ST-Link/VX JTAG tool.

## 3.16 Summary: Low Cost to Get Started

The chapter covered an exhaustive list of software tools and examples to download and install. What shouldn't be lost is that many of these tools are free. The financial barriers to entry are just a small cost for development boards and your time to learn and develop a solution. Before we jump into a project, we need to cover some STM32 Cortex-M basics.

# 4  STM32 Cortex Primer

The different projects are built with pre-developed code such as the HAL layers from STM32 libraries and the ThreadX Components. It is easy to forget that there are many little details to get the MCU up and running. For all the tools and examples downloaded in the last chapter, the datasheets and application notes that go with each STM32 MCU and development board are just as important. The details needed to go into bringing each MCU are beyond the scope of this book, but it is important to provide some high-level coverage so you know what to reference as you are designing and building your project. This chapter introduces the important details of the architecture, memory map, startup options, and clocks.

## 4.1  Different MCUs Different Architectures

There are many STM32 MCUs available. Each offers a different set of I/O that can fit a specific application. Microsoft chose three STM32 MCUs on which to base their ThreadX examples. These Cortex-M4 MCUs fit a wide range of applications but not every application, which became one of this book's driving questions. All the STM32 MCUs follow a generic architecture, which is presented in the diagram below.

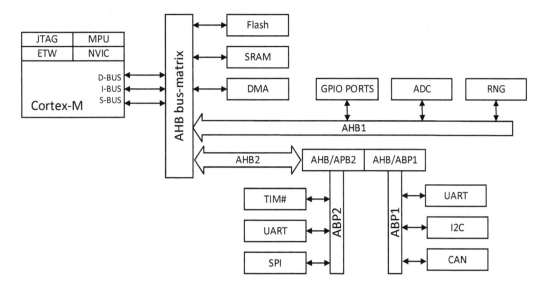

The MCU is centered on the Advanced High-performance Bus (AHB) Matrix that interconnects the master components such as the Cortex-M core and DMA with the slave components such as the Flash, SRAM, and the AHB to Advanced Peripheral Bus (APB). Fast-access memory devices such as flash, SRAM, and video memory are connected to the AHB busses directly. Slower serial I/O such as UART, I2C, and SPI are connected to the APB busses. The diagram is only an example, and you should refer to the MCU datasheet for the exact MCU you have chosen.

## 4.2  Memory Map

All peripherals connected to the busses and CPU are accessed via memory addressing. The MCU memory map and the address locations of each device are important. All STM32 MCUs have 4GB of memory address space (0x0000 0000 to 0xFFFF FFFF), which is divided into 8 sections. The following is the memory map for the STM32L4Rxxx and STM32L4Sxxx MCUs. Since every MCU has a slightly different memory map, please see the specific MCU programming guide for the memory map details of the device you have chosen.

| Section | Address | Contents |
|---|---|---|
| 7 | 0xFFFF FFFF ... 0xE000 0000 | Cortex-M / FPU / Internal Peripherals |
| 6 | 0xC000 0000 | Reserved |
| 5 | 0xA000 0000 | OCTOSPI |
| 4 | 0x8000 0000 | Flash Memory Controller / OCTOSPI |
| 3 | 0x6000 0000 | Flash Memory Controller |
| 2 | 0x4000 0000 | Peripherals (AHB / ABP) |
| 1 | 0x2000 0000 | SRAM/ GFX MMU |
| 0 | 0x0000 0000 | CODE |

The Code section is divided into different memory types. The first section is the boot memory section with the address starting at 0x0000 0000. What memory is aliased to this section is based on the BOOT configuration.

```
0x1FFF FFFF ┌─────────────────────┐
            │                     │
            │      Reserved       │
            │                     │
            ├─────────────────────┤
            │    Option bytes     │
            ├─────────────────────┤
            │      Reserved       │
            ├─────────────────────┤
            │      OTP area       │
            ├─────────────────────┤
            │   System Memory     │
            ├─────────────────────┤
            │      Reserved       │
            ├─────────────────────┤
            │       SRAM          │
            ├─────────────────────┤
            │      Reserved       │
            ├─────────────────────┤
            │    Flash memory     │
            ├─────────────────────┤
            │      Reserved       │
            ├─────────────────────┤
            │ Boot memory space   │
            │ (Flash, System      │
            │ Memory, or SRAM     │
            │ depends on BOOT     │
            │ options)            │
0x0000 0000 └─────────────────────┘
```

Embedded flash memory is split between regular flash memory and system memory. Since each STM32 MCU and development board has different peripherals and memory options, you will need to consult the datasheet and application notes for your specific details.

## 4.3  Start-Up

The boot mode on startup is already configured by the example boards. The STM32CubeProg can be used to configure boot settings, but we won't be addressing this capability in any of the projects. If you are going to build your own hardware, then understanding the different boot options is important. The boot options are determined by a combination of an external pin (BOOT0) and preprogrammed register bits (nSWBOOT0 and nBOOT# in FLASH_OPTR registers). Based on the combination of bits, the system can boot to Flash, SRAM, or system memory. The MCU datasheets have a table that lists the different combinations and what memory device is aliased to the boot memory space (0x0000 0000).

When the boot option is selected the memory device is aliased (remapped) to the boot memory space (0x0000 0000), but the memory device is still accessible in the original address location. For example, flash memory is located at 0x0800 0000. If the system is configured to boot from the Flash memory, the flash memory is aliased to the boot memory space (0x0000 0000), but it is still accessible from the original location of 0x0800 0000.

SRAM is located at 0x2000 0000. If the system is configured to boot from SRAM, the SRAM memory is aliased to the boot memory space (0x0000 0000), but it is still accessible from the original location of 0x2000 0000.

## 4.4 Clocks and Timers

There are a number of clock and timer features in each MCU. The system clock (SYSCLOCK) is the critical clock that affects the timing of all peripherals in the system. There are four sources for the system clock:

1. High-speed external crystal or ceramic resonator (HSE) oscillator, ranging from 4 to 48 MHz.
2. 16 MHz high-speed RC oscillator HSI16).
3. Multispeed internal RC oscillator (MSI).
4. System PLL which is fed by the HSE.

STM32 tools present the clock tree with all the clocks to peripherals and buses. You may adjust the clocks to fit your application power and performance requirements. The tools will flag any errors. You can either manually change the values or let the tools do this for you.

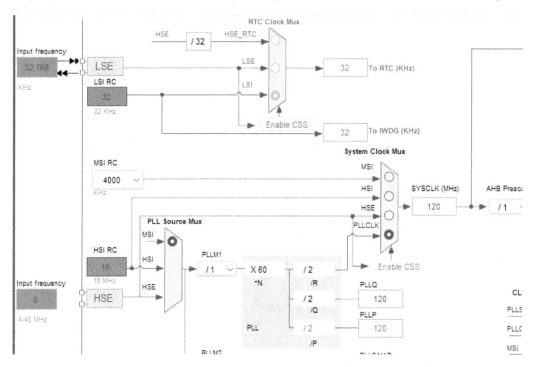

There are individual timers (TIM#) for advanced control, general purpose, basic, low-power, watchdog, and SysTick. The STM32 HAL code is configured to SysTick by default. Azure RTOS will want to take control of the SysTick, so the STM32 HAL time base needs to be changed to one of the basic or general-purpose timers (TIM#). This is a step that needs to be performed in the Azure RTOS project.

## 4.5 Development Kit Documentation

As mentioned in Chapter 2, STMicroelectronics has an extensive list of development boards and kits that can be used to start a project. Documentation for each of these kits is very important so you know where devices are placed in the memory map and how the clocks are configured by default.

## 4.6 Summary: It is all in the Details

The chapter provided a high-level view of the STM32 MCU architecture, memory map, start-up, and clocks. The details for each STM32 MCU will be different, so please consult the datasheet for the STM32 MCU that you will be using. With the tools installed and some high-level MCU architecture information covered, let's jump right into creating a project.

# 5   Project 1: Blinking LED Application No RTOS

Before we dive into ThreadX (Azure RTOS), the first project will help us get familiar with the tools and see the initialization of the startup code for the STM32 processor. The project is the famous Hello World for hardware, which is blink an LED, and we will implement two different versions of this project so you can see how to create a project from the processor only and then create a project using the full reference design. The STM32L4S5 Discovery Kit (B-L4S5I-IOT01A) will be the board used for this project. You may select a different board; but if you do, keep in mind the steps will be different.

## 5.1   Part 1: Initiate a Blinking LED Project 1 with STM32CubeMX

STM32CubeMX will be used to create the initialization code for the project. There are three paths to go down. The first is to select the actual MCU/MPU and define all the pins individually. The second is to select from a Development Kit Board. The third is to select an STM32 MCU/MPU from another silicon vendor's offerings. In this part, we will start with the MCU/MPU.

1.   Open STM32CubeMX.
2.   The open window allows you to select an existing project, or create a new project. Under the New Project, you start from a CPU/MCU, development board, or a sample application. Click on "Access to MCU Selector".
3.   The STM32L4S5 Discovery Kit (B-L4S5I-IOT01A) uses the STM32L4S5 VIT6 MCU, in the Commercial Part Number box enter STM32L4S5.
4.   A list of different package types appears on the right with information about the MCU, as well as, links to documentation. Select STM32L4S5VIT6 and click on Start Project.

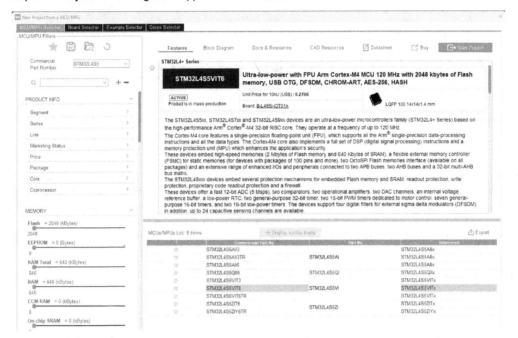

The project configuration opens to the Pinout & Configuration. If you click on Clock Configuration, Project Manager, and Tools tabs, you can see that there is a lot to configure at the very start of the project.

5.  In the Pinout & Configuration tab, the chip, with all the available pins, is displayed. Click on PB14. A context menu appears showing all the available pin options that can be set. Keep in mind that there will be conflicts with other I/O pins as each pin selection is made. Select GPIO_Output.

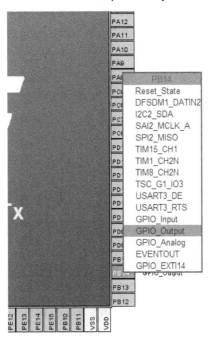

6.  Right-click on PB14 and select Enter user label.
7.  Enter: LED2 [LED_GREEN].

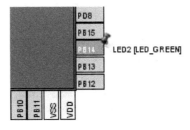

8.  On the left, under Categories, expand System Core and select GPIO. You will see the pin configuration settings.

9.  Click on the Project Manager tab.

10. Fill in the following:
    a. Project Name: Blink-LED-NoRTOS-NB
    b. Project Location: Make the project location the same location for the STM32CubeIDE Workspace folder.
    c. Toolchain/IDE: STM32CubeIDE.
    d. Click on the "Generate Under Root" checkbox.

11. Click on "Generate Code" in the top right. The project will be created in the folder you selected.
12. Once the code has been generated, a dialog will ask you to open the files or open the project. Click on the "Open Project" button.
13. STM32CubeIDE will open and import the project into the workspace folder that you created in Chapter 3.
14. Close STM32CubeMX.

## 5.2  Part 1: Project Set up in STM32Cube IDE

With STM32CubeIDE opened, expand the branches of the project on the left. All the initialization code for the I/O and memory has already been added and configured. The only thing that needs to be done is to write the application in the main.c file.

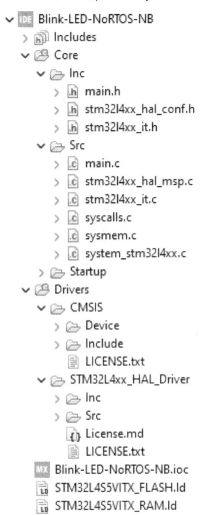

Before we jump into the code, let's cover the project architecture. The STM32L4S5VITX_FLASH.ld is the linker script that contains the memory segment definitions where the code will be placed in memory into the .elf output file. The startup_stm32l4s5vits.s file is the actual start-up that handles the reset vectors on startup, and it uses the linker script file to set up all the data segments and .bss global data such as the interrupt vectors for all I/O. The main function in main.c is called in the end. The main function first makes a call down to the primary HAL file to stm32i4xx_hal.c to reset all peripherals, initialize the flash interface, and the Systick (or system tick). The system and peripheral clocks are set up with a call to SystemClock_Config() and PeriphCommonClock_Config(). Next to last is a call to initialize all the peripherals that were set up in STM32CubeMX. Finally, we get to the Infinite loop which will run our code.

The files covered here are the basic setup files for all STM32 projects. When ThreadX (Azure RTOS) code is added, many more files will be added, and the startup process will end with a call into ThreadX.

## 5.3 Part 1: Modify the Main.c to Blink the LED with the HAL Functions

Now, we will add the code to main.c to blink the LED.

1. Open the main.c file. The main.c file contains all the calls to the startup code to initialize all the peripherals and memory. The file has many comments that show what code you can enter and that mark the locations where the code can be entered.
2. Under int main(void) is the start of the code. The HAL_Init() function is called to reset everything. Next, the system and peripheral clocks are configured. Finally, there are multiple calls to initialize the various I/O peripherals on the board. After all these calls is a while-loop where you can enter your code. Add the following code in the while-loop to toggle the LED on and off:

```
MX_USART3_UART_Init();
MX_USB_OTG_FS_USB_Init();
/* USER CODE BEGIN 2 */

/* USER CODE END 2 */

/* Infinite loop */
/* USER CODE BEGIN WHILE */
while (1)
{
        HAL_GPIO_TogglePin(LED2_GPIO_Port, LED2_Pin);
        HAL_Delay(1000);
    /* USER CODE END WHILE */

    /* USER CODE BEGIN 3 */
}
    /* USER CODE END 3 */
}
```

Since Toggling a GPIO is a common occurrence for this type of device, the STM32 HAL (stm32l4xxhal_gpio.c) provides the HAL_GPIO_TogglePin function to perform the toggling for you. The HAL_Delay() function (stm32l4xx_hal.c) delays the process before continuing with the loop.

**Warning**: It is very important to keep your code between the "User Code" comment sections. If you have to go back to the .ioc file to make changes, the code will be regenerated for the whole project. Any custom code inside the "User Code" comment sections will remain, but anything outside will be deleted.

3. Save the file.
4. Right-click on Blink-LED-NoRTOS-NB and select Build Project or hit CTRL+B. Correct any errors, but the build should complete successfully.

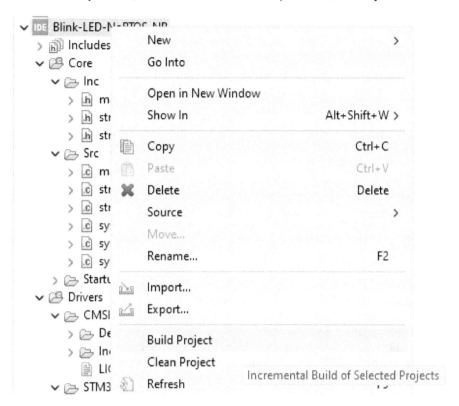

```
Console  X   Problems   Tasks   Properties
CDT Build Console [Blink-LED-NoRTOS-NB]

arm-none-eabi-size   Blink-LED-NoRTOS-NB.elf
arm-none-eabi-objdump -h -S  Blink-LED-NoRTOS-NB.elf  > "Blink-LED-NoRTOS-NB.list"
   text    data    bss    dec    hex filename
   6704      20   1572   8296   2068 Blink-LED-NoRTOS-NB.elf
Finished building: default.size.stdout

Finished building: Blink-LED-NoRTOS-NB.list


20:00:42 Build Finished. 0 errors, 0 warnings. (took 1s.756ms)
```

Build Analyzer  X   Static Stack Analyzer   Cyclomatic Complexity   Call Hierarchy   Type Hierarchy

Blink-LED-NoRTOS-NB.elf - /Blink-LED-NoRTOS-NB/Debug - Jul 27, 2023, 8:00:42 PM

Memory Regions | Memory Details

| Region | Start address | End address | Size | Free | Used | Usage (%) |
|---|---|---|---|---|---|---|
| RAM | 0x20000000 | 0x2009ffff | 640 KB | 638.45 KB | 1.55 KB | 0.24% |
| RAM2 | 0x10000000 | 0x1000ffff | 64 KB | 64 KB | 0 B | 0.00% |
| RAM3 | 0x20040000 | 0x2009ffff | 384 KB | 384 KB | 0 B | 0.00% |
| FLASH | 0x08000000 | 0x081fffff | 2 MB | 1.99 MB | 6.57 KB | 0.32% |

5. Before we debug the project on the board, we want to look at how the linker file aligns with the datasheet. Download the application note: RM0432, from the STMicroelectronics website.

6. In the RM0432 application note, look at the Figure 3 memory map. You will notice three sections for SRAM and the location for Flash memory.

7. Open the STM32L4S5VITX_FLASH.ld file.

8. In the Build Analyzer details, click on the "Memory Details" tab.

📇 Build Analyzer  ✕    ⚞ Static Stack Analyzer    ⓒ Cyclomatic Complexity    ⭏ Call Hierarchy    ⭏ Type Hierarchy

**Blink-LED-NoRTOS-NB.elf** - /Blink-LED-NoRTOS-NB/Debug - Jul 27, 2023, 8:00:42 PM

Memory Regions | Memory Details

Search

| Name | Run address (VMA) | Load address (LMA) | Size |
|------|-------------------|--------------------|------|
| ∨ ▦ FLASH | 0x08000000 | | 2 MB |
|   > ⭏ .isr_vector | 0x08000000 | 0x08000000 | 444 B |
|   > ⭏ .text | 0x080001bc | 0x080001bc | 6.05 KB |
|   > ⭏ .rodata | 0x080019f0 | 0x080019f0 | 64 B |
|   ⭏ .preinit_array | 0x08001a30 | 0x08001a30 | 0 B |
|   > ⭏ .init_array | 0x08001a30 | 0x08001a30 | 4 B |
|   > ⭏ .fini_array | 0x08001a34 | 0x08001a34 | 4 B |
|   > ⭏ .data | 0x20000000 | 0x08001a38 | 12 B |
| ▦ RAM2 | 0x10000000 | | 64 KB |
| ∨ ▦ RAM | 0x20000000 | | 640 KB |
|   ∨ ⭏ .data | 0x20000000 | 0x08001a38 | 12 B |
|     ▪ SystemCoreClock | 0x20000000 | 0x08001a38 | 4 B |
|     ▪ uwTickPrio | 0x20000004 | 0x08001a3c | 4 B |
|     ▪ uwTickFreq | 0x20000008 | 0x08001a40 | 1 B |
|   ∨ ⭏ .bss | 0x2000000c | | 32 B |
|     ▪ uwTick | 0x20000028 | | 4 B |
|   ⭏ ._user_heap_stack | 0x2000002c | | 1.5 KB |
| ▦ RAM3 | 0x20040000 | | 384 KB |

The memory details reflect what is in the linker description file, and provide the actual addresses for each section. We can see from the memory map in the data sheet that there are three SRAM sections. The linker description file shows SRAM1 and SRAM3 to be at their original addresses. SRAM2 is aliased to the boot memory address of 0x1000 0000. Flash is addressed to the boot memory section of 0x0800 0000. Now, look at the sizes. The total SRAM in the MCU is 640K. Per the datasheet, SRAM1 is 192 Kbytes, SRAM2 is 64Kbytes, and SRAM3 is 384 Kbytes. In the linker description file, SRAM1 length takes up the whole SRAM section.

```
/* Memories definition */
MEMORY
{
  RAM     (xrw)    : ORIGIN = 0x20000000,   LENGTH = 640K
  RAM2    (xrw)    : ORIGIN = 0x10000000,   LENGTH = 64K
  RAM3    (xrw)    : ORIGIN = 0x20040000,   LENGTH = 384K
  FLASH   (rx)     : ORIGIN = 0x8000000,    LENGTH = 2048K
}
```

The flash memory is broken into two distinct physical blocks:

- The main flash memory block is used for the application program and any user data.
- The information block contains three sub-parts:
    - Option bytes for hardware and memory protection user configuration.
    - System memory that contains ST proprietary code.
    - One-time programmable (OTP) area.

The total flash available is 2MB. Keep in mind that the flash is broken down into two different blocks. The build output files will be placed in the Debug folder of the project. The .elf file is the binary image to be programmed into the board.

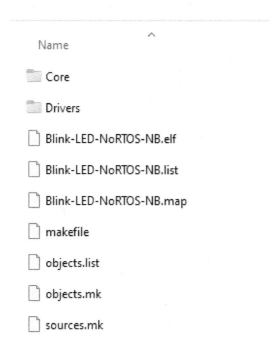

The list file disassembles the whole project C code to assembly and how the code is mapped to the different linker sections. The map provides a file and function view of what is in the .elf file.

## 5.4   Part 1: Debug the Application on the Board

We will now debug the application on the target hardware.

1. Connect the STM32L4S5 Discovery Kit's USB-B STLINK port to the development system via a USB-A to USB-B cable.

2. In STM32CubeICE, click the debug button 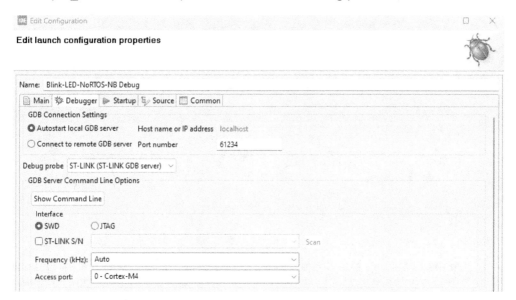 in the toolbar.
3. An Edit configuration will appear. Click on the Debugger tab and ST_LINK (ST_LINK GDB Server) should be set as the debug probe.

**Edit Configuration**

**Edit launch configuration properties**

Name:  Blink-LED-NoRTOS-NB Debug

Main  Debugger  Startup  Source  Common

GDB Connection Settings
- Autostart local GDB server      Host name or IP address  localhost
- Connect to remote GDB server  Port number           61234

Debug probe  ST-LINK (ST-LINK GDB server)  ∨
GDB Server Command Line Options

Show Command Line

Interface
- SWD            JTAG

- ST-LINK S/N                                    Scan

Frequency (kHz):  Auto
Access port:   0 - Cortex-M4

4. Click OK to start debugging. The efi file will be downloaded to the board. You may get a dialog asking to change to the debug perspective. Click yes.

**Note**: If you are asked to update the bootloader firmware, click Yes and follow the new window to install the new bootloader firmware. Once the update is complete, start the debug session, again.

5. The code will run and stop at HAL_init() in main.c. Click the Resume or hit the F8 key, and the green LED should be blinking on the board.
6. Stop debugging when finished. The code will continue to run and the LED will continue to flash on and off.
7. Close the main.c file in the STM32CubeIDE.

## 5.5  Part 2: Initiate the Blinking LED Project 2 with STM32CubeMX

Starting a project from the MCU is great for a custom board, but when you are evaluating an MCU, you will most likely choose one of the development boards available. Let's create our main project for the STM32L4S5 Discovery Kit (B-L4S5I-IOT01A).

1. Open STM32CubeMX.
2. The open window allows you to select an existing project, or create a new project. Under the New Project, you start from a CPU/MCU, development board, or a sample application. Click on "Access to Board Selector".

3. A new window appears. From the Commercial Part Number drop-down, select B-L4S5I-IOT01A. If you are using a different STM32 board, then select that product number.

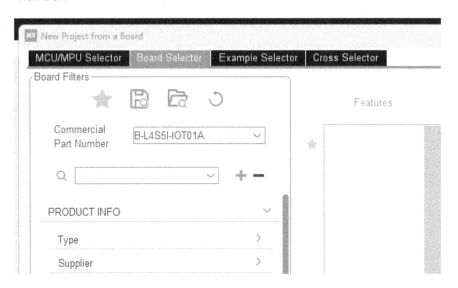

4. On the right-hand side, there is a list of boards associated with the part number. There is only one, so click on the item. The pane above files shows the information about the board. From here you can click on the links to get more resources for the board and the STM32L4+ MCU on the board.

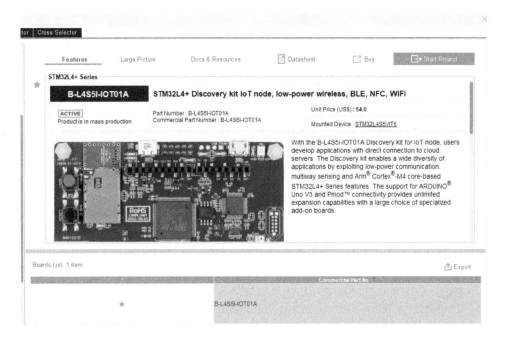

5. Click on "Start Project" in the top right corner.
6. You will be asked to initialize the default settings. Click Yes.
7. The project gets initialized and a picture of the MCU appears with all the pins and associated I/O configuration for the development board already laid out. If this were a custom board, you would go through all the steps with the test project to define each pin.

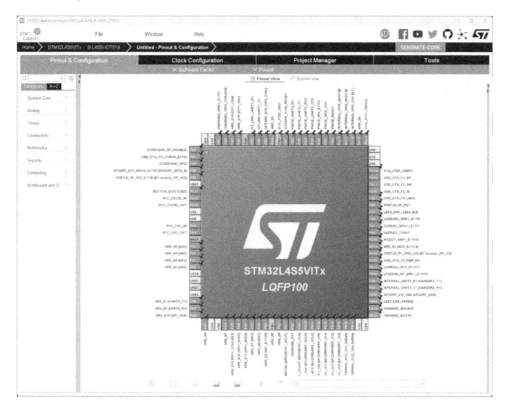

Again, the LED that will be toggled is LED2 [LED_GREEN] on Pin PN14 in the lower right corner.

8. Click on the Project Manager tab.
9. Fill in the following:
    a. Project Name: Blink-LED-NoRTOS.
    b. Project Location: Make the project location the same location as the STM32CubeIDE Workspace folder.
    c. Toolchain/IDE: STM32CubeIDE.
    d. Click on the "Generate Under Root" checkbox.

10. Click on "Generate Code" in the top right. The project will be created in the folder you selected.
11. Once the code has been generated, a dialog will ask you to open the files or open the project. Click on the "Open Project" button.
12. STM32CubeIDE will open and import the project into the workspace folder that you created in Chapter 3.
13. Close STM32CubeMX.

**Note**: A new project can be started in STM32CubeIDE, but the windowing of the setup is a bit constrained. It is not as obvious since we are using a development board with all the connections already defined. STM32CubeMX provides a cleaner more focused workspace to configure the chip the way you want it for your circuit.

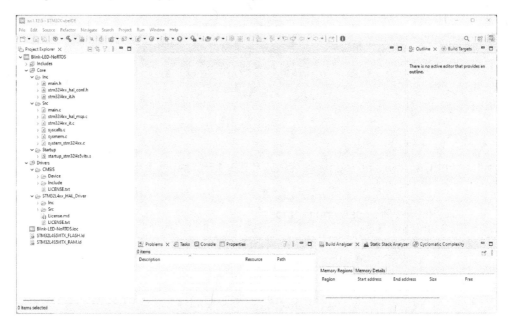

## 5.6   Part 2: Project Setup in STM32CubeIDE

With STM32CubeIDE opened, expand the branches under the Drivers\STM32L4xx_HAL_Driver\Src. Since we are basing the project on the full

44

development kit and selected to initialize all the I/O, there are more HAL drivers in this project, since it includes all of the development board I/O and features.

## 5.7 Part 2: Modify Main.c to Blink an LED Using HAL Functions

Now that we know how the system starts up, let's add our code to blink an LED.

1. Double-click on the Blink-LED-NoRTOS.ioc file.
2. A dialog appears asking if you want to open the Device Configuration tool in the Device Configuration Tool perspective. Click Yes.

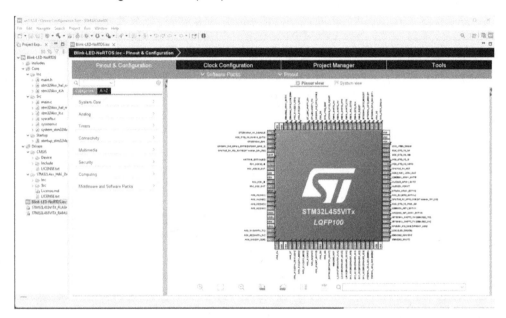

3. A tab will open and present the same interface that you saw in STM32CubeMX. From here you can make modifications and regenerate the code without having to go back to STM32CubeMX.
4. Close the Blink-LED-NoRTOS.inc tab.
5. Open the main.c file. The main.c file contains all the calls to the startup code to initialize all the peripherals and memory. The file has many comments that show what code you can enter and that mark the locations where the code can be entered.
6. Under int main(void) is the start of the code. The HAL_Init() function is called to reset everything. Next, the system and peripheral clocks are configured. Finally, there are multiple calls to initialize the various I/O peripherals on the board. After all these calls is a while-loop where you can enter your own code. Add the following code in the while-loop to toggle the LED:

```
MX_USART3_UART_Init();
MX_USB_OTG_FS_USB_Init();
/* USER CODE BEGIN 2 */

/* USER CODE END 2 */

/* Infinite loop */
```

```
/* USER CODE BEGIN WHILE */
while (1)
{
        HAL_GPIO_TogglePin(LED2_GPIO_Port, LED2_Pin);
        HAL_Delay(1000);
    /* USER CODE END WHILE */

    /* USER CODE BEGIN 3 */
}
/* USER CODE END 3 */
}
```

**Warning**: It is very important to keep your code between the "User Code" comment sections. If you have to go back to the .ioc file to make changes, the code will be regenerated for the whole project. Any custom code inside the "User Code" comment sections will remain, but anything outside will be deleted.

7. Save the file.
8. Right-click on Blink-LED-NoRTOS and select Build Project or hit CTRL+B. Correct any errors, but the build should complete successfully.

The project should build without errors.

The build output files will be placed in the Debug folder of the project. The .elf file is the binary image to be programmed into the board.

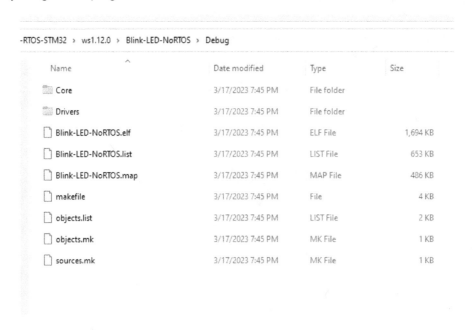

## 5.8   Part 2: Debug the Applications on the Board

We will now debug the application on the target hardware.

1. Connect the STM32L4S5 Discovery Kit's USB-B STLINK port to the development system via a USB-A to USB-B cable.

2. In STM32CubeICE, click the debug button                     in the toolbar.
3. An Edit configuration will appear. Click on the Debugger tab and ST_LINK (ST_LINK GDB Server) should be set as the debug probe.

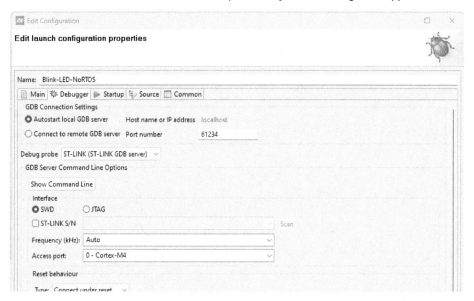

4. Click OK to start debugging. The efi file will be downloaded to the board. You may get a dialog asking to change to the debug perspective. Click yes.

**Note**: If you are asked to update the bootloader firmware, click Yes and follow the new window to install the new bootloader firmware. Once the update is complete, start the debug session, again.

5. The code will run and stop at HAL_init() in main.c. Click the Resume or hit the F8 key. The green LED should be blinking on the board.
6. Stop debugging when finished. The code will continue to run and the LED will continue to flash on and off.
7. Close the main.c file in the STM32CubeIDE.

## 5.9  Summary: First Look at the Tools

The goal of this project was to get familiar with the STM32 software development tools to create a project from scratch. We created two different projects, one starting from the MCU and the other from the development board. Much of the work to perform STM32 MCU initialization is automatically generated for you, so you can focus on your application. Having full access to the source code allows you to tweak and debug the code when necessary. Now that we know what the initialization code looks like, the next project will add the ThreadX kernel.

# 6   Project 2: ThreadX Blinking LED Application

Project 1 introduced the STM32 software development tools to create and deploy an application. For this project, we will create an ThreadX(Azure RTOS) application to toggle LED applications. Yes, same application, but this time the toggling with be in an ThreadX thread. We will see how ThreadX code is integrated with the initialization code and how it is launched on startup.

## *6.1   Initiate the Project with STM32CubeMX*

STM32CubeMX will be used to create the initialization code for the project.

1. Open STM32CubeMX.
2. The open window allows you to select an existing project, or create a new project. Under the new project, you start from a CPU/MCU, development board, or a sample application. Click on "Access to Board Selector".
3. A new window appears. From the Commercial Part Number drop-down, select B-L4S5I-IOT01A. If you are using a different STM32 board, then select that product number.

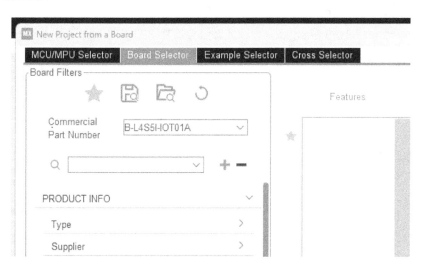

4. On the right-hand side, there is a list of boards associated with the part number. There is only one, click on the item. The pane above files shows the information

about the board. From here you can click on the links to get more resources about the board and STM32L4+ MCU on the board.

5. Click on "Start Project" in the top right corner.
6. You will be asked to initialize the default settings. Click Yes.
8. The project gets initialized and a picture of the MCU appears with all the pins and associated I/O configuration for the development board already laid out. If you were starting from an MCU or processor, you would start defining what each pin would be set to. Since this is a development board, we will keep the defaults. The LED that will be toggled is LED2 [LED_GREEN] on Pin PN14 in the lower right corner. If you click on Click Configuration, Project Manager, and Tools tabs, you can see there is a lot to configure at the very start of the project.

9. Now, we need to add ThreadX(Azure RTOS) to the project. Click on Software Packs->Select Components:

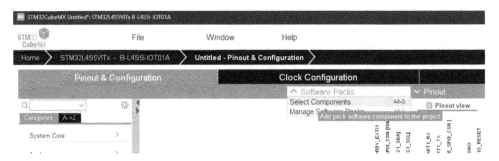

10. The package selector appears. There are some packages active and some that are inactive. The active packages are for the STM32 MCU on the board. The inactive packages are for other STM32 MCUs. Locate STMicroelectronics.X-CUBE-AZRTOS-L4 and click the Install button next to the package. This will install the Azure RTOS package support for the STM32 MCU that is on the board.
11. Once installed, expand the branches under RTOS ThreadX->ThreadX.
12. Tick the box next to Core.

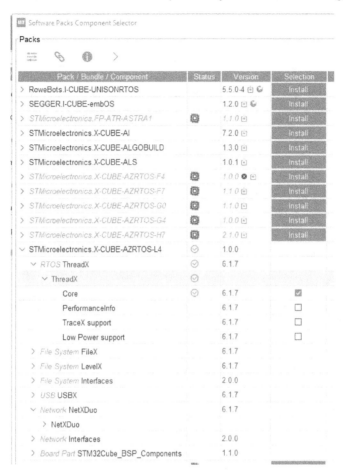

13. Click the OK button in the bottom right corner.
14. STMicroelectronics.X-CUBE-AZRTOS-L4 is now listed under Middleware and Software Packs. Click on STMicroelectronics.X-CUBE-AZRTOS-L4:

15. The STMicroelectronics.X-CUBE-AZRTOS-L4 Mode and Configuration will appear. Check the RTOS ThreadX. The configuration options are displayed below the Mode. We will keep the defaults.
16. We need to configure the clock. Expand the System Core on the left side.
17. The SYS-tick is used by the HAL and Azure RTOS. To separate the two, we will give the HAL a different clock source. In the categories, select System Core->SYS.
18. In the Timebase Source drop-down, select TIM6.

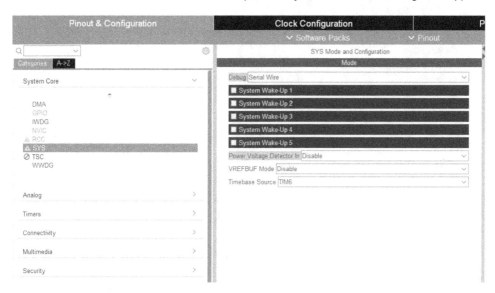

18. Click on the Project Manager tab.
19. Fill in the following:
    a. Project Name: AzureRTOSBlink-LED
    b. Project Location: you may choose the folder path for the MX project.
    c. Toolchain/IDE: STM32CubeIDE.
    d. Click on the "Generate Under Root" checkbox.

20. Click on Generate Code. The project will be created in the folder you selected.
21. Once the code has been generated, a dialog will ask you to open the files or open the project. Click on the "Open Project" button.

22. STM32CubeIDE will open and import the project into the workspace folder that you created in Chapter 3.
23. Close STM32CubeMX.

## 6.2   ThreadX File Additions

We see the same startup files as the previous project. The additional ThreadX (AzureRTOS) Software Package added several new source and header files to the initialization code from the previous project. The whole ThreadX library is added under the Middlewares\ST directory. The project contains an Azure RTOS application. The AZURE_RTOS folder can be used for setting up ThreadX.

Open main.c, and scroll down to main(void). The same start-up initialization calls take place. There is a new call to MX_ThreadX_Init() that kicks off the AzureRTOS kernel and never returns.

```
MX_USART1_UART_Init();
MX_USART2_UART_Init();
MX_USART3_UART_Init();
MX_USB_OTG_FS_USB_Init();
/* USER CODE BEGIN 2 */

/* USER CODE END 2 */

MX_ThreadX_Init();

/* We should never get here as control is now taken by the
scheduler */
/* Infinite loop */
/* USER CODE BEGIN WHILE */
while (1)
{
    /* USER CODE END WHILE */

    /* USER CODE BEGIN 3 */
}
/* USER CODE END 3 */
}
```

MX_ThreadX_Init() is found in app_threadx.c. Open app_threadx.c, and the function simply calls  tx_kernel_enter(), which is a define for the _tx_initialize_kernel_enter(void) function found in _tx_initialize_kernel_enter.c, which in turn can be found in the \Middlewares\ST\threadX\common\src folder. The \Middlewares\ST\threadX\common\src folder contains the rest of the AzureRTOS Core functions and files, of which there are many. STM32CubeIDE allows you to perform the familiar search on function using a right-click so you can easily navigate around the code to see the source code behind each function call. Our intent with the book is to keep things at a high level to focus on the functions and features, but you are more than welcome to dig deeper into the code. Eventually, the _tx_initialize_kernel_enter() function makes a call to tx_application_define() in app_azure_rtos.c, which sets up a memory pool for ThreadX and a call to App_ThreadX_Init(), which is back in app_threadx.c. The app_threadx.c file is where we spend most of the time creating the application.

## 6.3   Edit the code

After all the additions, we only need to edit the app_threadx.c.file to blink the LED.

1. In STM32CubeIDE, double-click on app_threadx.c to open the file for editing.

2. At about line 21 in the include section, add the main.h and stdio.h include statements. Main.h provides the definitions for the LED port, and stdio.h provides the standard C printf call that will be used in the application.

```
/* USER CODE END Header */

/* Includes ------------------------------------------------------------
---------------*/
#include "app_threadx.h"
#include "main.h"
#include <stdio.h>
```

**Note**: these two includes should be placed in the "User Code Beging Includes" block, but I want to demonstrate something in the next chapter.

3. At line 37, enter the following code to define the thread stack size to 1024.:

```
/* Private define ------------------------------------------------------
---------------*/
/* USER CODE BEGIN PD */
#define THREAD_STACK_SIZE 1024
/* USER CODE END PD */
```

4. At line 47, fill in the code to define a byte array based on stack size and a thread pointer.

```
/* Private variables ---------------------------------------------------
---------------*/
/* USER CODE BEGIN PV */
uint8_t thread_stack[THREAD_STACK_SIZE];
TX_THREAD thread_ptr;
/* USER CODE END PV */
```

5. At line 53, declare the custom thread entry to be called in the private function prototypes.

```
/* Private function prototypes -----------------------------------------
---------------*/
/* USER CODE BEGIN PFP */
VOID my_thread_entry(ULONG intial_input);
/* USER CODE END PFP */
```

6. Now add the call to create the thread in APP_ThreadX_Init() function.

```
UINT App_ThreadX_Init(VOID *memory_ptr)
{
    UINT ret = TX_SUCCESS;
    TX_BYTE_POOL *byte_pool = (TX_BYTE_POOL*)memory_ptr;

    /* USER CODE BEGIN App_ThreadX_Init */
    (void)byte_pool;
    tx_thread_create(&thread_ptr, "led_thread", my_thread_entry,
0x1234, thread_stack, THREAD_STACK_SIZE, 15,15,1,TX_AUTO_START);
    /* USER CODE END App_ThreadX_Init */

    return ret;
}
```

The tx_thread_create function has multiple parameters. There are a couple entries to point out. The first 15 value is the priority level, and 1 is the time slice.

7. Finally, we can add the my_thread_entry to blink the LED in the USER CODE BEGIN 1 section at the end of the file.

```
/* USER CODE BEGIN 1 */
VOID my_thread_entry(ULONG initial_input){
        printf("Thread Entry Reached\n");
        printf("\n");


        printf("Starting Loop...\n");
        while(1){
                HAL_GPIO_TogglePin(LED2_GPIO_Port, LED2_Pin);
/*Defined in main.h*/
                printf("hello\n");
                tx_thread_sleep(200);
        }
}
```

8. Save the app_threadx.c file.

The printf() function will send a string out of the UART port(USB STLINK). Per a STM32 application note, the C library function needs to be made for messages to be sent out the UART port.

9. Open main.c
10. At line 907, add the following code to the "USER CODE BEGIN 4" block:

```
/* USER CODE BEGIN 4 */
/**
  * @brief  Retargets the C library printf function to the UART.
```

```
 * @param   None
 * @retval None
 */
PUTCHAR_PROTOTYPE
{
  /* Place your implementation of fputc here */
  /* e.g. write a character to the USART1 and Loop until the end
of transmission */
  HAL_UART_Transmit(&huart1, (uint8_t *)&ch, 1, 0xFFFF);

  return ch;
}
/* USER CODE END 4 */
```

11. Save the file. You will notice errors flagged. We need to make one final change.
12. Open main.h
13. Add the following define the USE CODE private defines:

```
/* USER CODE BEGIN Private defines */
#define PUTCHAR_PROTOTYPE int __io_putchar(int ch)
/* USER CODE END Private defines */
```

14. Save the file.
15. Right-click on AzureRTOSBlink-LED and select Build Project or hit CTRL+B. Correct any errors, but the build should complete successfully.

```
CDT Build Console [AzureRTOSBlink-LED]
arm-none-eabi-gcc -o "AzureRTOSBlink-LED.elf" @"objects.list"   -mcpu=cortex-m4
Finished building target: AzureRTOSBlink-LED.elf

arm-none-eabi-size   AzureRTOSBlink-LED.elf
arm-none-eabi-objdump -h -S AzureRTOSBlink-LED.elf  > "AzureRTOSBlink-LED.list"
   text    data     bss     dec     hex filename
  36052     124    6764   42940    a7bc AzureRTOSBlink-LED.elf
Finished building: default.size.stdout

Finished building: AzureRTOSBlink-LED.list


20:49:01 Build Finished. 0 errors, 0 warnings. (took 8s.875ms)
```

## 6.4   Debug the Applications on the Board

With the application built, it is time to deploy the AzureRTOSBlink-LED.elf file to the hardware and test the application.

1. In app_threadx.c, set a breakpoint at the first printf() call in the my_thread_entry() function.
2. Connect the STM32L4S5 Discovery Kit's USB-B STLINK port to the development system via a USB-A to USB-B cable.

3. In STM32CubeICE click the debug button  from the tool bar.
4. An Edit configuration will appear. Click on the Debugger tab and ST_LINK (ST_LINK GDB Server) should be set as the debug probe.
5. Click Ok to start debugging. The efi file will be downloaded to the board. You may get a dialog asking to change to the debug perspective. Click yes.
6. The code will run and stop at HAL_init() in main.c. Click the Resume or hit he F8 key. The debugger will hit the breakpoint.
7. Step through the code and you can watch the LED turn on and off.
8. Stop debugging when finished. The code will continue to run.
9. Open an Annabooks COM Terminal or equivalent serial terminal application.
10. Set the COM port to the STM32L4S5 Discovery Kit's USB COM port and initiate a connection.
11. Hit the reset button, and watch the output terminal window display.

12. Stop the terminal application when finished.

## 6.5   Add a Second Thread

Now let's add a second thread to the application, so we can see the interaction between the threads.

1. Open App_threadx.c.
2. Around line 47, add a new thread stack and pointer after the original thread.

```
/* Private variables ------------------------------------------
--------------*/
/* USER CODE BEGIN PV */
uint8_t thread_stack[THREAD_STACK_SIZE];
TX_THREAD thread_ptr;
uint8_t thread_stack2[THREAD_STACK_SIZE];
TX_THREAD thread_ptr2;
/* USER CODE END PV */
```

3. In the section at about Line 55, add the function prototype for the thread after the first thread.

```
/* Private function prototypes ----------------------------------
--------------*/
/* USER CODE BEGIN PFP */
VOID my_thread_entry(ULONG intial_input);
VOID my_thread_entry2(ULONG intial_input);
/* USER CODE END PFP */
```

4. In App_ThreadX_Init() function. add the call to start the thread.

```
UINT App_ThreadX_Init(VOID *memory_ptr)
{
  UINT ret = TX_SUCCESS;
  TX_BYTE_POOL *byte_pool = (TX_BYTE_POOL*)memory_ptr;

  /* USER CODE BEGIN App_ThreadX_Init */
  (void)byte_pool;
  tx_thread_create(&thread_ptr, "led_thread", my_thread_entry,
0x1234, thread_stack, THREAD_STACK_SIZE, 15,15,1,TX_AUTO_START);
  tx_thread_create(&thread_ptr2, "second_thread",
my_thread_entry2, 0x2468, thread_stack2, THREAD_STACK_SIZE,
14,14,1,TX_AUTO_START);
  /* USER CODE END App_ThreadX_Init */

  return ret;
```

```
}
```

5. Finally, add the thread at the end of the file after the first thread that toggles the LED.

```
VOID my_thread_entry2(ULONG initial_input){
        printf("Thread2 Entry Reached\n");
        printf("\n");


        printf("Starting Loop...\n");
        while(1){
                printf("Thread2\n");
                tx_thread_sleep(150);
        }
}
/* USER CODE END 1 */
```

6. Save the file.
7. Right-click on AzureRTOSBlink-LED and select Build Project or hit CTRL+B. Correct any errors, but the build should complete successfully.
8. Make sure the STM32L4S5 Discovery Kit is connected to the development machine via the USB cable and start the debug session.
9. When the breakpoint is hit, click continue.
10. Stop Debugging
11. Open an Annabooks COM Terminal or equivalent serial terminal application.
12. Set the COM port to the STM32L4S5 Discovery Kit's USB COM port and initiate a connection.
13. Hit the reset button, and watch the output terminal window display.

You see the output messages from both threads. Since the first thread has a longer sleep time, the second thread will run twice before the first thread runs again.

14. Stop the terminal application when finished.


## 6.6   Summary: First ThreadX Application

Blinking an LED gets a little more involved when it is an ThreadX application. The project's main goal here was to show the ThreadX core files that get added to the STM32 project and how the initialization code jumps to ThreadX with a single function call. We learned how to set up a thread and retarget the printf() library function to output to the debug UART port. Now that we have covered the basics, we can dive into more sophisticated applications and explore a diagnostic feature.

# 7 Project 3: Threads and TraceX

TraceX provides a post-mortem analysis of the sequence of threads that ran during a debug session. It allows you to see the sequential order that the threads ran along with the elapsed time that the threads were running. With this trace information, you can adjust the thread priority levels to meet your system's design requirements. We will add TraceX support to the last project that we developed in the previous chapter, and show how to use TraceX to tune the thread sequencing.

## 7.1 Add The TraceX Software Package

1. Make sure that you have downloaded the TraceX application from the Microsoft store.
2. With the AzureRTOSBlink-LED project opened in STM32CubeIDE, open the AzureRTOSBlink-LED.ioc file. This will open the configuration.
3. Click Software Packs->Software Components.
4. Expand STMicroelectronics.X-CUBE-AZTRTOS-L4.
5. Expand the branches under RTOS ThreadX until you see the TraceX support.
6. Check the box next to TraceX support.
7. Click Ok.

8. On the left side of STM32CubeIDE, expand Middleware and Software Packs.
9. Click on X-CUBE-AZRTOS-L4. This will open the Mode and Configuration pane.

10. Under configuration, expand Trace, and enable TX_ENABLE_EVENT_TRACE.

11. Finally, regenerate the code to add TraceX to the build. From the menu, select Project->Generate Code, or click on the yellow gear from the toolbar.
12. When asked to change perspective, click Yes.

Under the Middlewares->ST->thread->common-src, you will see the newly added trace files that have been added to the project:

```
> .c  tx_timer_system_activate.c
> .c  tx_timer_system_deactivate.c
> .c  tx_timer_thread_entry.c
> .c  tx_trace_buffer_full_notify.c
> .c  tx_trace_disable.c
> .c  tx_trace_enable.c
> .c  tx_trace_event_filter.c
> .c  tx_trace_event_unfilter.c
> .c  tx_trace_initialize.c
> .c  tx_trace_interrupt_control.c
> .c  tx_trace_isr_enter_insert.c
> .c  tx_trace_isr_exit_insert.c
> .c  tx_trace_object_register.c
> .c  tx_trace_object_unregister.c
> .c  tx_trace_user_event_insert.c
> .c  txe_block_allocate.c
```

Adding TraceX requires some changes to be made to the original code to take advantage of the TraceX features. We will make these changes in the next section.

## 7.2   Edit the Code to Add the Trace Buffer

Within the code, we need to define the Trace buffer where the raw TraceX data will be stored.

1. Open STM32L4S5VITX_FLASH.id, which defines how memory is allocated.
2. Scroll down until you see the Memories definition. This is how RAM and flash are configured on the board. If you have a different board, the memory mapping will be a little different. We are going to use part of the 640K RAM to store the trace buffer.

```
/* Memories definition */
MEMORY
{
  RAM      (xrw)    : ORIGIN = 0x20000000,   LENGTH = 640K
  RAM2     (xrw)    : ORIGIN = 0x10000000,   LENGTH = 64K
  RAM3     (xrw)    : ORIGIN = 0x20040000,   LENGTH = 384K
  FLASH    (rx)     : ORIGIN = 0x8000000,    LENGTH = 2048K
}
```

3. After the Memories definition comes the SECTIONS, which defines what is in each memory. Scroll down to just after .Fini_array  and add the .trace section:

```
.fini_array :
{
  . = ALIGN(4);
  PROVIDE_HIDDEN ( __fini_array_start = .);
  KEEP (*(SORT(.fini_array.*)))
  KEEP (*(.fini_array*))
  PROVIDE_HIDDEN ( __fini_array_end = .);
  . = ALIGN(4);
} >FLASH

.trace (NOLOAD):
{
  . = ALIGN(4);
  *(.trace)
} >RAM

/* Used by the startup to initialize data */
```

4. Save and close the file.
5. Open app_threadx.c.
6. Scroll down to the Private variables and add the buffer size after the thread sizes.

```
/* Private variables -------------------------------------------
--------------*/
/* USER CODE BEGIN PV */
uint8_t thread_stack[THREAD_STACK_SIZE];
TX_THREAD thread_ptr;
uint8_t thread_stack2[THREAD_STACK_SIZE];
TX_THREAD thread_ptr2;
#define TRACEX_BUFFER_SIZE 64000
uint8_t tracex_buffer[64000];
/* USER CODE END PV */
```

7.  In the App_ThreadX_Init() function, add the code to enable tracing before starting the threads.

```
UINT App_ThreadX_Init(VOID *memory_ptr)
{
  UINT ret = TX_SUCCESS;
  TX_BYTE_POOL *byte_pool = (TX_BYTE_POOL*)memory_ptr;

  /* USER CODE BEGIN App_ThreadX_Init */
  (void)byte_pool;
  tx_trace_enable(&tracex_buffer,TRACEX_BUFFER_SIZE,30);
  tx_thread_create(&thread_ptr, "led_thread", my_thread_entry,
0x1234, thread_stack, THREAD_STACK_SIZE, 15,15,1,TX_AUTO_START);
  tx_thread_create(&thread_ptr2, "second_thread",
my_thread_entry2, 0x2468, thread_stack2, THREAD_STACK_SIZE,
14,14,1,TX_AUTO_START);
  /* USER CODE END App_ThreadX_Init */

  return ret;
}
```

8.  In the previous chapter, you were directed to place the includes outside the User Code blocks. Since the code was modified when the TraceX software package was added, we have to re-add some code. Scroll up to the includes and add the missing includes, but this time the includes will be added to the correct user code block:

```
/* Includes ---------------------------------------------------
--------------*/
#include "app_threadx.h"

/* Private includes -------------------------------------------
--------------*/
/* USER CODE BEGIN Includes */
#include "main.h"
#include <stdio.h>
```

```
/* USER CODE END Includes */
```

9. Save the file.
10. Right-click on AzureRTOSBlink-LED and select Build Project or hit CTRL+B. Correct any errors, but the build should complete successfully.

## 7.3   Debug the Application and Capture the Trace Buffer

Now, we will run the debugger to generate the trace.

1. Make sure the STM32L4S5 Discovery Kit is connected to the development machine via the USB cable and start the debug session.
2. The debug will stop at HAL_Init(). Click continue to run the debugger a bit, and then stop debugging.
3. Open the Memory view. From the menu, select Windows->Show view->Memory.
4. In the Memory view, click on the Plus (+) symbol so we can add the tracex_buffer variable to monitor.

Once you click OK. You will see the start of memory where the trace buffer begins.

**Note**: any misspelling of the buffer variable will result in an error message.

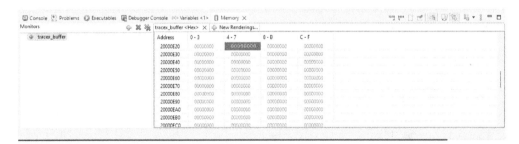

5. Click continue in the debugger.
6. Let the application run for a bit to fill up the buffer, and then suspend or pause the debugging. DO NOT STOP THE DEBUGGER. You will see that the buffer is now filled with data.

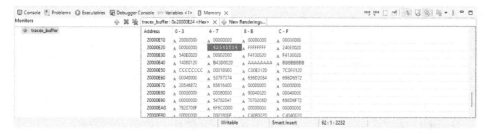

7. Click on the export button so we can export the data to a file that can be read with TraceX.
8. A dialog opens, set the following:
    a. Format: RAW Binary.
    b. Length: 6400.
    c. File Name: <folder location>\<file name>.trx

**Note**: Make sure to give the file name a .trx extension. You can put the file in the TraceX folder under the project.

9. Click OK to export the buffer to the file.

For future reference, you can add the file association for .trx to TraceX.

10. From the menu, select Windows->Preferences.
11. Expand General->Editors.
12. Click on File Associations.
13. Click on the Add button next to File Types.
14. Enter .trx and click Ok.
15. Click on the new .trx entry and click on the Add button next to Associated editors.
16. A dialog appears for Editor Selection. Check the External program radio button.
17. Enter tracex in the search, and select TraceX Trace File.

18. Click Ok.
19. Click Apply and Close.

## 7.4   View Buffer Data in TraceX

We can now view the buffer data in TraceX.

1.   Open TraceX.
2.   Open the .trx file.

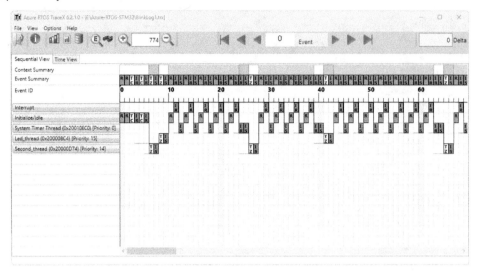

The data is opened in the Sequential view, and you can see the transitions being made from the different states and the threads. If you collected enough data, you should see that the thread2 ran twice before the LED ran in some instances because of the time. The results will match what we saw earlier. In the menu, you can open the ThreadX legend to get the timing:

3.  Click on the Time view tab.
4.  Zoom out to see more detail on the thread activity over time. You will have to zoom out to see any transitions. For the most part, the threads are small and run very quickly, keeping the system in the idle state most of the time.

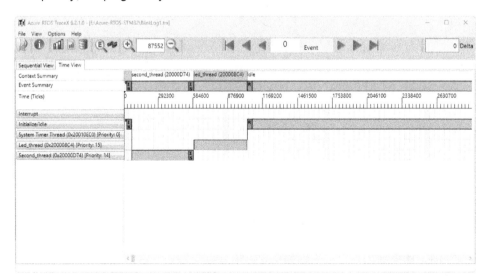

5.  Close TraceX when finished.

## 7.5 Summary: A View into Application Running

TraceX is a post-mortem debug analysis too that can be used to get a big picture of the thread and interrupt activity that is going on while an application is running. Live debugging, while the system is running, provides a more real-time approach, TraceX offers a comprehensive view of the relationship of all of the system threads that a live debug session would be challenged to catch.

# 8   Project 4: Barometer -No RTOS

So far, we have created projects using software packages and just added a few lines of code. In Chapter 3, we downloaded an embedded firmware package that contained source code for device drivers, sample projects, etc. This project will show where software packages like this can come up short and how to add external files to a project as needed. The project will output the data from the temperature, humidity, and pressure sensors that are on the STM32LS4S5 Discovery Kit board to the serial port.

## 8.1   Initiate the Project with STM32CubeMX

Let's create a new project that only uses the HAL code. The project will not include the Azure RTOS package to simplify the explanation.

1.   Open STM32CubeMX.
2.   Under the New Project, click on "Access to Board Selector".
3.   A new window appears. From the Commercial Part Number drop-down, select B-L4S5I-IOT01A. If you are using a different STM32 board, then select that product number.

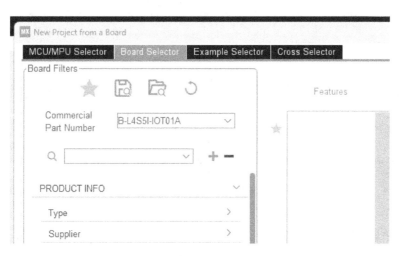

4.   On the right-hand side, there is a list of boards associated with the part number. There is only one, click on the item. The pane above files shows the information about the board. From here you can click on the links to get more resources about the board and STM32L4+ MCU on the board.

5. Click on "Start Project" in the top right corner.
6. You will be asked to initialize the default settings. Click Yes.
7. The project gets initialized and a picture of the MCU appears with all the pins and associated I/O configuration for the development board already laid out. Since this is a development board, we will keep the defaults.
8. Click on the Project Manager tab.
9. Fill in the following:
    a. Project Name: Barometer-NoRTOS
    b. Project Location: Make the project location the same location as the STM32CubeIDE Workspace folder.
    c. Toolchain/IDE: STM32CubeIDE.
    d. Click on the "Generate Under Root" checkbox.

| Project Settings | |
|---|---|
| Project Name | Barometer-NoRTOS |
| Project Location | E:\Azure-RTOS-STM32\ws1.12.0\ |
| Application Structure | Advanced |
| Toolchain Folder Location | \Azure-RTOS-STM32\1.12.0\Barometer-NoRTOS\ |
| Toolchain / IDE | STM32CubeIDE |

10. Click on "Generate Code" in the top right. The project will be created in the folder you selected.
11. Once the code has been generated, a dialog will ask you to open the files or open the project. Click on the "Open Project" button.
12. STM32CubeIDE will open and import the project into the workspace folder that you created in Chapter 3.
13. Close STM32CubeMX.

## 8.2   MEMS Drivers Versus Embedded Firmware Package Drivers

The ThreadX Getting Started Example (GSE) has a complete solution to send sensor date to Azure IoT Central. The running question is "how did someone create this project and how can we replicate from scratch?" We are going to look at the differences between two drivers sets and the working project drivers to see which path is best to achieve the goal. The humidity temperature sensor HTS221 and the pressure sensor LPS22HB are sensors are to the I2C2 bus port. Please see the board schematic for more information.

### 8.2.1   Add the MEMS1 Drivers
Let's add the drivers from the software packages.

1. Open STM32CubeIDE, if it is not already open.
2. Under the Barometer-NoRTOS project, open the Barometer-NoRTOS.ioc file.

3.  The familiar STM32CubeMX project file appears. Click on Software Packs->Select Components.
4.  Install the STMIcroelectronics.X-CUBE-MEMS1 package, if it is not already installed.
5.  Expand the STMIcroelectronics.X-CUBE-MEMS1 branch. You will see various supported sensors.
6.  Expand Board Part HumTemp.
7.  In the selection column for HTS221, select I2C from the drop-down.
8.  In the selection column for the LPS22HB, select I2C from the drop-down.

| | | | |
|---|---|---|---|
| ⌄ STMicroelectronics.X-CUBE-MEMS1 | ⊘ | 9.5.0 ⌄ | |
| › Exposed APIs | | | |
| › Device MEMS1_Applications | | 9.5.0 | |
| › Board Part AccGyr | | 5.5.0 | |
| › Board Part AccMag | | 5.6.0 | |
| › Board Part Acc | | 1.4.0 | |
| Board Part AccTemp / LIS2DTW12 | | | Not selected ⌄ |
| › Board Part Mag | | 5.4.0 | |
| ⌄ Board Part HumTemp | ⊘ | 5.5.0 | |
| HTS221 | ⊘ | 5.5.0 | I2C ⌄ |
| SHT40AD1B | | | Not selected ⌄ |
| ⌄ Board Part PressTemp | ⊘ | 5.5.0 | |
| LPS22HB | ⊘ | 5.5.0 | I2C ⌄ |
| LPS22HH | | | Not selected ⌄ |
| LPS33HW | | | Not selected ⌄ |
| LPS33K | | | Not selected ⌄ |

9.  Click OK to close the Select Components page.
10. Now we need to configure the drivers. Under Categories, expand Middleware and Software Packs.
11. Click on X-CUBE-MEMS1.
12. Check the boxes for both sensor parts.
13. Set the Found Solutions to I2C2 for both parts.

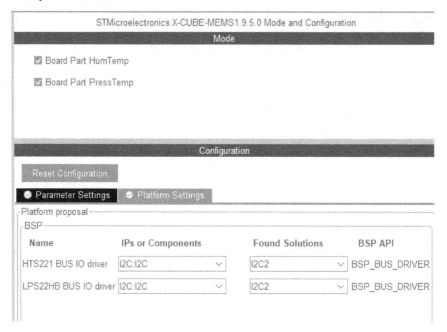

14. From the menu, select Project->Generate Code. The source code for the sensors will be added to the project.

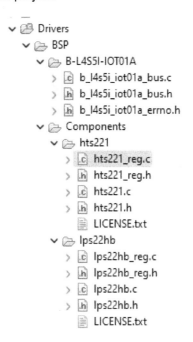

### 8.2.2    Compare Source Code

Using WinMerge, we will compare the HTS221 driver sets:

1.   Open WinMerge.
2.   Click on the Folder icon to do a folder compare.
3.   Open the first folder to the \Barometer-NoRTOS\Drivers\BSP\Components\hts221.
4.   Open the second folder to \STMicroelectronics\B-L4S5I-IOT01A\lib\stm32cubel4\ Drivers\BSP\Components\hts221.
5.   Click Compare, and WinMerge shows that the drivers are very different. The MEMS1 driver is broken into 4 files, whereas the getting started driver is only two files.

6.   Open the hts221.c and you will see that the drivers are completely different. In fact, they were developed by two different teams.

7.   Close the hts221.c comparison.
8.   Close the comparison results.
9.   Change Folder 1 to the hts221 driver in the embedded firmware package: C:\Users\<user                              account>\STM32Cube\Repository \STM32Cube_FW_L4_V1.17.2\ Drivers\BSP\Components\hts221
10.  Click Compare, and you will see that the results are identical.

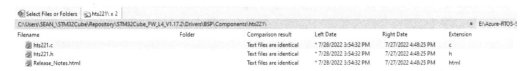

The drivers for the working project came from the embedded firmware package. A closer examination of the files shows that the embedded firmware package driver has full implementation and simpler API for use with the main application. The call to read the temperature, humidity, and pressure is a simple function call that returns a floating-point value. All the complex registry calculations that are required to read the data on the sensors are already performed in the driver source code. The MEMS1 drivers are generic so they fit with the tools and flexibility of selecting the BUS ports.

### 8.2.3    Remove the MEMS1 Drivers

To add the correct drivers, we need to remove the MEMS1 drivers.

1. In STM32CubeIDE open the Barometer-NoRTOS.ioc file, if not already open.
2. Under Categories, expand Middleware and Software Packs.
3. Click on X-CUBE-MEMS1.
4. Uncheck the boxes for both sensor parts.
5. From the menu, select Project->Generate Code. The code for the sensors will be removed from the project.
6. Close the Barometer-NoRTOS.ioc file.

### 8.2.4    Add the Embedded Firmware Drivers

Now, we are going to add the necessary source code files from the embedded firmware package to the project.

1. Open two file explorer windows.
2. In the first explorer window open the location of the embedded firmware package: C:\Users\<user account>\STM32Cube\Repository\STM32Cube_FW_L4_V1.17.2\Drivers\BSP.
3. In the second explorer window, open the location of the project: \Barometer-NoRTOS\Drivers.
4. In the second explorer window for the project, create a folder called BSP in \Barometer-NoRTOS\Drivers.
5. In the first window copy over the B-L4S5I-IOT01 folder to the \Barometer-NoRTOS\Drivers\BSP folder.
6. In the newly copied B-L4S5I-IOT01 folder in the project, remove all the files except the following:

| Name | Date modified | Type | Size |
|---|---|---|---|
| stm32l4s5i_iot01.c | 7/28/2022 3:54 PM | C Source | 25 KB |
| stm32l4s5i_iot01.h | 7/28/2022 3:54 PM | C/C++ Header | 10 KB |
| stm32l4s5i_iot01_hsensor.c | 7/28/2022 3:54 PM | C Source | 3 KB |
| stm32l4s5i_iot01_hsensor.h | 7/28/2022 3:54 PM | C/C++ Header | 3 KB |
| stm32l4s5i_iot01_psensor.c | 7/28/2022 3:54 PM | C Source | 3 KB |
| stm32l4s5i_iot01_psensor.h | 7/28/2022 3:54 PM | C/C++ Header | 3 KB |
| stm32l4s5i_iot01_tsensor.c | 7/28/2022 3:54 PM | C Source | 3 KB |
| stm32l4s5i_iot01_tsensor.h | 7/28/2022 3:54 PM | C/C++ Header | 3 KB |

These are all the files for the humidity, temperature, pressure, and bus drivers. They add the API that the application will use to access the sensor information.

7.  Go back to the first window and copy the Components folder to the \Barometer-NoRTOS\Drivers\BSP folder.
8.  In the newly copied Components folder in the project, remove all the folders except the following:

| Name | Date modified | Type |
|---|---|---|
| Common | 4/22/2023 8:53 PM | File folder |
| hts221 | 4/22/2023 8:53 PM | File folder |
| lps22hb | 4/22/2023 8:53 PM | File folder |

These are the driver and header files.

9.  Close both file explorer windows.
10. Go back to STM32CubeIDEIn the Project Explorer, right-click on the project, and select Refresh. The driver BSP folder with the driver files will appear.
11. Next, we need to configure the project to include these files in the build. From the menu, select Project->Properties.
12. Select C/C++ Build->Settings.
13. Expand the MCU GCC Compiler and select include paths.

14. Click on the Add symbol        .
15. Click on the Workspace button.
16. Open the \Barometer-NoRTOS\Drivers\BSP\ B-L4S5I-IOT01 folder.

81

17. Click OK.
18. Click OK, again.
19. Repeat steps 14-18 to add the \Barometer-NoRTOS\Drivers\BSP\Components\hts221 and \Barometer-NoRTOS\Drivers\BSP\Compnents\lps22hb folders.

20. Click Apply and Close.
21. A dialog will appear asking about changes that will take place during the rebuild. Click Yes.

**Note**: These last few steps are the long way to add the include-file paths. It is important to see where the settings are stored. The simpler solution is to right-click on the folder and select "Add/remove include path…" from the context menu.

## 8.3   Edit the Code

With the driver files in place and the project configured to include these files into the build, we can now work on the application.

1.   In STM32CubeIDE, open the main.c file. There are going to be several edits, so we will walk through the file from top to bottom.
2.   Starting around line19, add the includes to the header files for the drivers and C function libraries:

```
/* USER CODE END Header */
/* Includes ------------------------------------------------
--------------*/
#include "main.h"
#include <stdio.h>


/* Private includes ----------------------------------------
--------------*/
/* USER CODE BEGIN Includes */
#include "stm32l4s5i_iot01.h"
#include "stm32l4s5i_iot01_tsensor.h"
```

```
#include "stm3214s5i_iot01_hsensor.h"
#include "stm3214s5i_iot01_psensor.h"
#include <math.h>
/* USER CODE END Includes */
```

3. We need to add the code to make Printf() use the serial port. We will add the prototype to the function around line 37:

```
/* Private define ------------------------------------------
---------------*/
/* USER CODE BEGIN PD */
#define PUTCHAR_PROTOTYPE int __io_putchar(int ch)
/* USER CODE END PD */
```

4. In the main() function after all the calls to initialize I/O, add the following code to User Code Begin 2:

```
/* USER CODE BEGIN 2 */
printf("Barometer values measurements\n\n\r");
printf("=====> Initialize HTS221 sensor  \r\n");
BSP_TSENSOR_Init();
printf("=====> HTS221 sensor  initialized \r\n ");
printf("=====> Initialize LPS22hb sensor \r\n");
BSP_PSENSOR_Init();
printf("=====> LPS22hb sensor  initialized \r\n ");
printf("=====> Initialize Humidity sensor \r\n");
BSP_HSENSOR_Init();
printf("=====> Humidity sensor  initialized \r\n ");
/* USER CODE END 2 */
```

5. In the while-loop, add the following code after User Code Begin 3:

```
HAL_GPIO_TogglePin(LED2_GPIO_Port, LED2_Pin);
HAL_Delay(1000);

//Get Temperature
float temp_value = roundf(BSP_TSENSOR_ReadTemp()*100)/100;
printf("Temperature (degC): %0.2f\n", temp_value);
HAL_Delay(1000);
float pres_value = roundf(BSP_PSENSOR_ReadPressure()*100)/100;
printf("Pressure (hPa): %0.2f\n", pres_value);
HAL_Delay(1000);
float humid_value = roundf(BSP_HSENSOR_ReadHumidity()*100)/100;
printf("Humidity (%%rH): %0.2f\n", humid_value);
```

There is going to be an error flagged about the use of floating-point. We will fix this after all the code has been added.

6.  The last code to add is for the PUTCHAR_PROTOTYPE. Scroll down to about line 879. In the User Code Begin 4 section add the following code:

```
/* USER CODE BEGIN 4 */
PUTCHAR_PROTOTYPE
{
  /* Place your implementation of fputc here */
  /* e.g. write a character to the USART1 and Loop until the end
of transmission */
  HAL_UART_Transmit(&huart1, (uint8_t *)&ch, 1, 0xFFFF);

  return ch;
}
/* USER CODE END 4 */
```

7.  Save the file.
8.  Now, let's go back and fix the floating-point issue. From the menu, select Project->Properties
9.  C/C++Build->Settings.
10. In the MCU Settings, check the box for "User float with printf from newlib-nano (-u _printf_float)".
11. Click Apply and Close. The error goes away.

12. Right-click on Barometer-NoRTOS and select Build Project or hit CTRL+B. Correct any errors, but the build should complete successfully.

## 8.4   Debug the Applications on the Board

We are ready to test the application on the board.

1.  Make sure the STM32L4S5 Discovery Kit is connected to the development machine via the USB cable and start the debug session.
2.  The debuger will stop at HAL_Init(). Click continue to run the debugger a bit, and then stop debugging.
3.  Open an Annabooks COM Terminal or equivalent serial terminal application.
4.  Set the COM port to the STM32L4S5 Discovery Kit's USB COM port and initiate a connection.
5.  Hit the reset button, and watch the output terminal window display. The sensor readings appear on the output of the terminal.

 Annabooks COM Terminal

```
Pressure (hPa): 999.40
Humidity (%rH): 42.11
Temperature (degC): 25.65
Pressure (hPa): 999.31
Barometer values measurements


=====> Initialize HTS221 sensor
=====> HTS221 sensor  initialized
 =====> Initialize LPS22hb sensor
=====> LPS22hb sensor  initialized
 =====> Initialize Humidity sensor

=====> Humidity sensor  initialized
 Temperature (degC): 25.71
Pressure (hPa): 999.35
Humidity (%rH): 42.35
Temperature (degC): 25.73
Pressure (hPa): 999.32
Humidity (%rH): 42.22
Temperature (degC): 25.75
Pressure (hPa): 999.22
Humidity (%rH): 42.13
Temperature (degC): 25.76
Pressure (hPa): 999.36
Humidity (%rH): 42.13
Temperature (degC): 25.75
Pressure (hPa): 999.37
Humidity (%rH): 42.01
Temperature (degC): 25.76
Pressure (hPa): 999.34
Humidity (%rH): 41.97
Temperature (degC): 25.75
Pressure (hPa): 999.46
```

6.  Close the terminal and STM32CubeIDE when finished.


## 8.5   Summary: Customizing the Project

The embedded firmware package offers a better solution than the generic software package. If you are building your own board and put the sensors on a different I2C bus, you will have to adjust the code for that all the same. The project also showed how to add code from the outside and configure the project to include the files when performing the build. This project helps us understand how the GSE and sample IDE projects from Eclipse ThreadX site were created.

# 9   Project 5: NetX Duo

We have covered 2 of the 7 cornerstones of ThreadX. In the last chapter, we covered the differences between the software and firmware packages, which led to integrating source code from an outside source. So far, we can create an ThreadX (Azure RTOS) project, and we can access the sensors of the STM32L4S5 Discovery Kit board. The next step is accessing the WiFi chip on the STM32L4S5 Discovery Kit. This brings us to the next ThreadX cornerstone: NetX Duo, which contains a complete TCP/IP software stack, as well as, network features such as SNTP, DNS, MQTT, PPP, DHCP Client, etc.

You have already seen the NetX Duo software package in the STM32Cube tools, but this package is already a version behind the sample code provided by Microsoft, as well as, the official NetX Duo available from GitHub.

This project will borrow from one of the sample projects from Microsoft to set up the WiFi so you can ping the STM32L4S5 Discovery Kit from a different computer.

## 9.1   Run the Example Project

The STM32CubeIDE samples on the Microsoft site contain a NetX Duo Ping project. Our first step is to build this project.

### 9.1.1   Import the Workspace Project

To import the project:

1. Make sure that you have downloaded the STM32CubeIDE Example project, which should have been completed back in Section 3.11.
2. Open STM32CubeIDE.
3. From the menu, select File->import.
4. Click on "Existing Projects into Workspace'.
5. Click Next.
6. Browse to the directory containing the stm32cubeide folder, and click on Select folder.
7. All the projects should be checked, and click Finish.

### 9.1.2   Build and Debug the Ping Project

With the example projects loaded, we can build and run the Ping project.

1. In the sample_netx_duo_ping project, expand the branches until you see board_setup.c.
2. Edit the board_setup.c file to add your WIFI_SSID and WIFI_PASSWORD.
3. Save the file.
4. Right-click on sample_netx_duo_ping.
5. Select "Build Project" from the context menu. The project should build without error.
6. Connect the STM32L4S5 Discovery Kit to the development computer using the USB cable.
7. Start the debug session. The debugger will deploy the .elf file to the board and reboot the board. The debugger will stop at the board_setup() function.
8. Start ABCOMTerm or a similar serial terminal program and connect to the board's COM port.
9. In STM32CubeIDE, click on the continue button ▶ to let the program run.

The terminal program will show the output from the STM32L4S5 Discovery Kit.

STM32L4XX Lib:
> CMSIS Device Version: 1.7.0.0.
> HAL Driver Version: 1.12.0.0.
> BSP Driver Version: 1.0.0.0.
ES-WIFI Firmware:
> Product Name: Inventek eS-WiFi
> Product ID: ISM43362-M3G-L44-SPI
> Firmware Version: C3.5.2.5.STM
> API Version: v3.5.2
ES-WIFI MAC Address: C4:7F:51:91:44:40
wifi connect try 1 times
ES-WIFI Connected.
> ES-WIFI IP Address: 192.168.1.41
> ES-WIFI Gateway Address: 192.168.1.1
> ES-WIFI DNS1 Address: 192.168.1.1
> ES-WIFI DNS2 Address: 8.8.8.8

10. With the successful WiFi connection, open a command prompt and ping the IP address of the STM32L4S5 Discovery Kit and you should get a reply.

C:\Users\AB>ping 192.168.1.41

Pinging 192.168.1.41 with 32 bytes of data:
Reply from 192.168.1.41: bytes=32 time=16ms TTL=255
Reply from 192.168.1.41: bytes=32 time=3ms TTL=255
Reply from 192.168.1.41: bytes=32 time=3ms TTL=255
Reply from 192.168.1.41: bytes=32 time=3ms TTL=255

Ping statistics for 192.168.1.41:
   Packets: Sent = 4, Received = 4, Lost = 0 (0% loss),
Approximate round trip times in milli-seconds:
   Minimum = 3ms, Maximum = 16ms, Average = 6ms

We now know that this WiFi project works. Let's recreate the project from scratch; but before we do, a little research is in order.

## 9.2 Example and Software Package Research

To save some time but still be educational, let's review the steps used to recreate this solution. WinMerge was extensively used to compare the directories of the example projects, software packages, and firmware packages to see which was the latest and most comprehensive code. There was a lot of investigation into the different samples to understand how the system starts up and enables the WiFi. There were little nuggets of discovery along the way.

The STM32Cube software packages contain a solution for NetX Duo. A test project was created that contained ThreadX and NetX Duo.

- One thing missing from the STM32CubeMX setup is a driver for the Inventek ISM43362 WiFi chip. An STM32Cube software package was not available for this chip, so the source code has to be drawn from the Microsoft example.
- A WinMerge comparison between the NetX Duo directory in the project and with Net Duo directory showed that the example projects had a newer version of NetX Duo.

STM32Cube software packages appear to be a nice starting point, but the software packages are not up to date with the latest release of ThreadX. Creating a new STM32 project will have to exclude the NetX Duo software package, and we will have to substitute the NetX Duo directory from one of the examples. We will have to be very careful and match exactly how the sample_netx_duo_ping project is set up. An attempt to move the NETX Duo directory ran into the following issues:

- Missing include paths.
- A folder closed in the sample_netx_duo_ping project was activated in this project. This resulted in missing the curl.h and cmocka.h tools in the Azure IoT SDK subdirectory. This is going to be a common theme for the Azure IoT SDK, which will be covered a couple more times in the book.

**Note**: An update during the writing of this book, brought the NetX Duo components in the STM32 tools up to the level of the examples. The latest solution can always be downloaded from GitHub directly.

As we work through creating the project, we will implement the solution to these problems.

The next step was to research the WiFi driver source code. A software package in the STM32Cube repository is not available, but the firmware package appears to have some sort of available driver. The driver would be for HAL-level code, which doesn't come with a TCP/IP stack. Looking into both example projects, there are multiple files that comprise a solution for the Inventek ISM43362 WiFi chip. These files are scattered in different locations in each project. The files will be gathered and placed in the BSP directory of our new project.

Finally, there is the method for launching the WiFi and NetX Duo. The first thing one notices about the example projects is they look nothing like the projects created with the STM32Cube tools. For this example, the main.c file is replaced with the sample_netx_duo_ping.c file that contains the main() function. This function calls a board_setup() function that is in board_setup.c, which performs all the hardware initializations. Starting a project from scratch will require pulling code out of sample_netx_duo_ping.c and board_setup.c so we can take advantage of the STM32Cube tools to set up the HAL layer for the project.

**Note**: This took a few days to work out the bugs before being able to replicate the Ping project.

## 9.3 NetX Duo Component Add-ons

NetX Duo is comprised of a core component, several add-on components, and security components. You can see this break-down when you look at the software packages.

| | | | | |
|---|---|---|---|---|
| ⊟ *Network* NetXDuo | ⊘ | 6.1.7 | | |
| ⊟ NetXDuo | ⊘ | | | |
| NX Core | ⊘ | 6.1.7 | ☑ | |
| TraceX Support | | 6.1.7 | ☐ | |
| Addons AutoIP | ⊘ | 6.1.7 | ☑ | |
| Addons DHCP Client | ⊘ | 6.1.7 | ☑ | |
| Addons DHCP Server | | 6.1.7 | ☐ | |
| Addons DNS | ⊘ | 6.1.7 | ☑ | |
| Addons mDNS | | 6.1.7 | ☐ | |
| Addons MQTT | ⊘ | 6.1.7 | ☑ | |
| Addons Cloud | ⊘ | 6.1.7 | ☑ | |
| Addons PPP | ⊘ | 6.1.7 | ☑ | |
| Addons SNTP | ⊘ | 6.1.7 | ☑ | |
| Addons Web Client | | 6.1.7 | ☐ | |
| Addons Web Server | | 6.1.7 | ☐ | |
| TLS | ⊘ | 6.1.7 | ☑ | |
| Crypto | ⊘ | 6.1.7 | ☑ | |

Each add-on package is a networking feature such as web server, DHCP Client, DNS, etc. Based on the selections above, the following source code directory structure gets created:

The Example Package with the Ping Example has the following source code directory

As you can see, a different set of add-on components were used. The latest NetX Duo downloaded from GitHUB has many more components and features to choose from.

Some add-ons conflict with other add-ons, and some add-ons need components like FileX to function. The details of NetX Duo architecture and components are online. What is important is that you can remove unnecessary functionality when building a custom solution, which is what was clearly done for the Microsoft Example Projects.

## 9.4   Initiate the Project with STM32CubeMX

Now, we are ready to create the project. The project will use the ThreadX software package in the STM32Cube repository. NetX Duo will be added to the project from the STM32CubeIDE example, and the build paths will be entered manually. Finally, code from sample_netx_duo_ping.c and board_setup.c of the Ping Example will be selectively reused in the application.

1.   Open STM32CubeMX.
2.   Under the New Project, click on "Access to Board Selector".
3.   A new window appears. From the Commercial Part Number drop-down, select B-L4S5I-IOT01A. If you are using a different STM32 board, then select that product number.
4.   On the right-hand side, there is a list of boards associated with the part number. In this case there is only one. Click on the item. The pane above files shows the information about the board. From here you can click on the links to get more resources about the board and STM32L4+ MCU on the board.
5.   Click on "Start Project" in the top right corner.

6. You will be asked to initialize the default settings. Click Yes. The project gets initialized and a picture of the MCU appears with all the pins and associated I/O configuration for the development board already laid out. Since this is a development board, we will keep the defaults.

7. Now, we need to add ThreadX (Azure RTOS) to the project. Click on Software Packs->Select Components

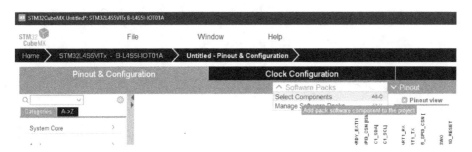

8. The package selector appears. Some packages are active and some are inactive. The active packages are for the STM32 MCU on the board. The inactive packages are for other STM32 MCUs. Locate STMicroelectronics.X-CUBE-AZRTOS-L4 and click the Install button next to the package. This will install the Azure RTOS package support for the STM32 MCU that is on the board.

9. Once installed, expand the branches under RTOS ThreadX->ThreadX.

10. Tick the box next to Core.

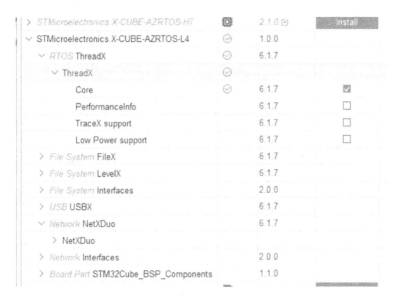

11. Click the OK button in the bottom right corner.

12. STMicroelectronics.X-CUBE-AZRTOS-L4 is now listed under Middleware and Software Packs. Click on STMicroelectronics.X-CUBE-AZRTOS-L4,

13. The STMicroelectronics.X-CUBE-AZRTOS-L4 Mode and Configuration will appear. Check the RTOS ThreadX. The configuration options are displayed below the Mode. We will keep the defaults.
14. We need to configure the clock. Expand the System Core on the left side.
15. The SYS-tick is used by the HAL and Azure RTOS. To separate the two, we will give the HAL a different clock source. In the categories, select System Core->SYS.
16. In the Timebase Source drop-down, select TIM6.

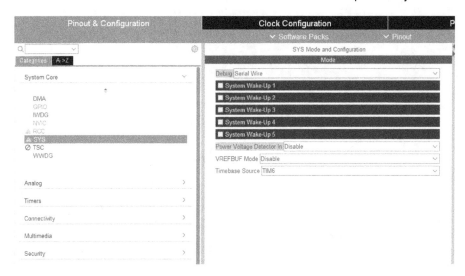

17. Finally, we need to add some HAL features to help with security. In the Categories tab, expand the Security section and enable AES, HASH, and RNG (Random Number Generator).

18. Click on the Project Manager tab.
19. Fill in the following:
    a.  Project Name: AzureRTOS-NetXDuo.
    b.  Project Location: Make the project location the same location as the STM32CubeIDE Workspace folder.
    c.  Toolchain/IDE: STM32CubeIDE.
    d.  Click on the "Generate Under Root" checkbox.

20. Click on "Generate Code" in the top right. The project will be created in the folder you selected.
21. Once the code has been generated, a dialog will ask you to open the files or open the project. Click on the "Open Project" button.
22. STM32CubeIDE will open and import the project into the workspace folder that you created in Chapter 3.
23. Close STM32CubeMX.

## 9.5   Add NetX Duo from STM32CubeIDE Examples

Integrating the NetX Duo source into a known working project involves some trimming and adding of build paths to the project settings.

1.  Open File Explorer.
2.  Open the location to the STM32CubeIDE example project.

RTOS-STM32  >  b-l4s5i-iot01a_2022_11_30  >  stm32cubeide

| Name | Date modified |
|---|---|
| common_hardware_code | 3/28/2023 8:14 PM |
| docs | 3/28/2023 8:14 PM |
| netxduo | 3/29/2023 8:20 PM |
| sample_azure_iot_embedded_sdk | 3/29/2023 8:18 PM |
| sample_azure_iot_embedded_sdk_pnp | 3/29/2023 8:18 PM |
| sample_netx_duo_ping | 3/29/2023 8:24 PM |
| sample_threadx | 3/29/2023 8:18 PM |
| stm32l4xx_lib | 3/29/2023 8:21 PM |
| threadx | 3/29/2023 8:22 PM |
| tools | 3/28/2023 8:15 PM |
| .project | 3/29/2023 8:18 PM |
| LICENSE.txt | 3/28/2023 8:14 PM |
| LICENSED-HARDWARE.txt | 3/28/2023 8:14 PM |
| NOTICE.txt | 3/28/2023 8:14 PM |
| SECURITY.txt | 3/28/2023 8:14 PM |
| TestA.txt | 3/30/2023 3:15 PM |

3.  Right-click on the netxduo folder and select copy.
4.  Under the new project, open the following folder:   \AzureRTOS-NetXDuo\Middlewares\ST.
5.  You will see the threadx folder that came from the STM32Cube responsory. Paste the netxduo folder to the directory. You will now have two folders threadx and netxduo.

**Note**: It might be better practice to put the netxduo on the same directory level as the ST folder since the source code didn't come from the STM32Cube repository. This is the developer's discretion. Be careful, if you have to go back to the STM32CubeMX interface and regenerate the code. The netxduo directory will disappear.

6. Now, we need to remove some unnecessary items. The Azure IOT SDK for C has a number of sample and test projects included. If these are left in place, the build will fail. In File Explorer, go to AzureRTOS-NetXDuo\Middlewares \ST\netxduo\addons\azure_iot\azure-sdk-for-c\sdk.
7. Delete the "samples" and "tests" folders.
8. Next delete the \AzureRTOS-NetXDuo2\Middlewares\ST\netxduo\addons\azure_iot \azure-sdk-for-c\sdk\src\azure\platform folder.

## 9.6   Add the Wi-Fi driver

The source code for the Inventek ISM43362 WiFi chip is scattered in the STM32CubeIDE Example Project.

1. Open File Explorer.
2. In the new project, create a BSP folder under \AzureRTOS-NetXDuo\Drivers.
3. Under \AzureRTOS-NetXDuo\Drivers, create a WiFi folder and a B-L4S5I-IOT01 folder.
4. In the STM32CubeIDE Example project folder, go to \stm32cubeide\stm32l4xx_lib.
5. Copy stm32l4s5i_iot01.c and stm32l4s5i_iot01.h and paste them to the \AzureRTOS-NetXDuo\Drivers\BSP\ B-L4S5I-IOT01 folder.
6. In the \stm32cubeide\stm32l4xx_lib folder, copy all the ex_wifi* files and paste them to the \AzureRTOS-NetXDuo\Drivers\BSP\WiFi folder.
7. In the \stm32cubeide\sample_netx_duo_ping folder, copy the nx_driver_stm32l4.c and nx_driver_stm32l4.h and paste them to the \AzureRTOS-NetXDuo\Drivers\BSP\WiFi folder.
8. In the stm32cubeide\common_hardware_code folder, copy es_wifi.c, wifi.c, and Wifi.h and paste them to the \AzureRTOS-NetXDuo\Drivers\BSP\WiFi folder.

The BSP subfolder folders should contain the following files,

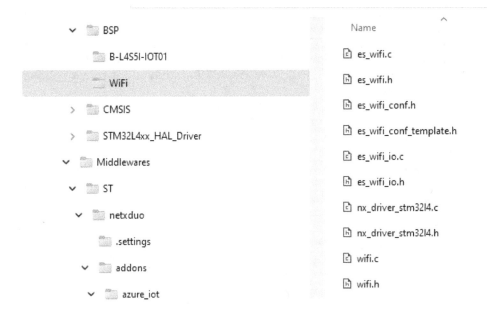

## 9.7   Set the Include Paths

Now we need to add the compiler build paths to the project.

1.   In STM32Cube IDE, click on the AzuireRTOS-NetXDuo project.
2.   From the menu, select Project->Properties.
3.   A Properties dialog appears. Expand the C/C++ Build and click on Settings.
4.   Go to MGU GCC Compiler->Include Paths.

You will see a number of include paths already set up for the HAL and for threadx. Now, we need to add all the paths for netx duo and the BSP folders. You could add each folder individually through this interface or you could simply right-click on the folder and select Add/Remote include path from the context menu.

5.   To make things a little easier, in the Chapter 8 Exercise that comes with the book, there is an IncludePaths.txt file. Open the file in Notepad or Notepad++. The file contains the following.

**../Core/Inc**
**../AZURE_RTOS/App**
**../Drivers/STM32L4xx_HAL_Driver/Inc**
**../Drivers/STM32L4xx_HAL_Driver/Inc/Legacy**
**../Drivers/CMSIS/Device/ST/STM32L4xx/Include**
**../Drivers/CMSIS/Include**
**../Middlewares/ST/threadx/common/inc/**
**../Middlewares/ST/threadx/ports/cortex_m4/gnu/inc/**

```
"${workspace_loc:/${ProjName}/Middlewares/ST/netxduo/addons/azure_iot/azure_iot_security_module/iot-security-module-core/deps/flatcc/include}"
"${workspace_loc:/${ProjName}/Middlewares/ST/netxduo/addons/azure_iot/azure_iot_security_module/iot-security-module-core/deps/flatcc/src/runtime}"
"${workspace_loc:/${ProjName}/Middlewares/ST/netxduo/addons/azure_iot/azure-sdk-for-c/sdk/inc}"
"${workspace_loc:/${ProjName}/Middlewares/ST/netxduo/addons/azure_iot}"
"${workspace_loc:/${ProjName}/Middlewares/ST/netxduo/ports/cortex_m4/gnu/inc}"
"${workspace_loc:/${ProjName}/Middlewares/ST/netxduo/nx_secure/inc}"
"${workspace_loc:/${ProjName}/Middlewares/ST/netxduo/nx_secure/ports}"
"${workspace_loc:/${ProjName}/Middlewares/ST/netxduo/crypto_libraries/inc}"
"${workspace_loc:/${ProjName}/Middlewares/ST/netxduo/common/inc}"
"${workspace_loc:/${ProjName}/Middlewares/ST/netxduo/addons/dns}"
"${workspace_loc:/${ProjName}/Middlewares/ST/netxduo/addons/mqtt}"
"${workspace_loc:/${ProjName}/Middlewares/ST/netxduo/addons/web}"
"${workspace_loc:/${ProjName}/Middlewares/ST/netxduo/addons/cloud}"
"${workspace_loc:/${ProjName}/Drivers/BSP/B-L4S5I-IOT01}"
"${workspace_loc:/${ProjName}/Drivers/BSP/WiFi}"
"${workspace_loc:/${ProjName}/Middlewares/ST/netxduo/addons/azure_iot/azure_iot_security_module}"
"${workspace_loc:/${ProjName}/Middlewares/ST/netxduo/addons/azure_iot/azure_iot_security_module/inc}"
"${workspace_loc:/${ProjName}/Middlewares/ST/netxduo/addons/azure_iot/azure_iot_security_module/inc/configs/RTOS_BASE}"
"${workspace_loc:/${ProjName}/Middlewares/ST/netxduo/addons/azure_iot/azure_iot_security_module/iot-security-module-core/inc}"
"${workspace_loc:/${ProjName}/Middlewares/ST/netxduo/addons/sntp}"
```

**Note**: It took some time to figure out the paths and removing the test code from the Azure IoT C SDK to get the system to build cleanly. This is a problem with the Azure IoT C SDK that will be covered in the last chapter.

6. Copy all the contents, and past the contents to the Include paths section by hitting a CTRL+V on the keyboard.

7. Click on the "Apply and Close" button.

## 9.8   Edit the Code

We have already seen how an Azure RTOS project is set up with the STM32Cube tools. The main.c file performs the I/O setup and then makes a call to start the kernel in app_threadx.c. For this project, there is going to be an extra step to initialize the Wi Fi driver and start up NetX Duo. We will pull source code from the STM32CubeIDE Example to build the application.

### 9.8.1   Create a Wi Fi Setup File

There are two files in the sample_netx_duo_ping project that define the application: sample_netx_duo_ping.c and board_setup.c. The board_setup.c file contains the initialization code for the hardware including the Wi-Fi. Our new project already contains the hardware setup with the exception of the Wi-Fi so we will pull out the Wi-Fi setup into a new file for our project. The sample_netx_duo_ping.c contains the main() function that calls board_setup and then starts the Azure RTOS kernel. The file also contains the information for configuring and initializing NetX Duo.  There is already a file that is in our project with the function call to initialize Azure RTOS, so we will put the code there.

1. In STM32CubeIDE, under \AzureRTOS-NetXDuo\Core\Inc, create a file called wifi_setup.h.
2. Open wifi_setup.h and enter the following code. Be sure to add your Wi-FI SSID, password, and WiFi security type.

```c
#ifndef __WIFISETUP_H
#define __WIFISETUP_H

/* Define the default wifi ssid and password. The user can
override this
   via -D command line option or via project settings.  */

#ifndef WIFI_SSID
//#error "Symbol WIFI_SSID must be defined."
#define WIFI_SSID ""
#endif /* WIFI_SSID  */

#ifndef WIFI_PASSWORD
//#error "Symbol WIFI_PASSWORD must be defined."
#define WIFI_PASSWORD ""
#endif /* WIFI_PASSWORD  */

/* WIFI Security type, the security types are defined in wifi.h.
   WIFI_ECN_OPEN = 0x00,
   WIFI_ECN_WEP = 0x01,
   WIFI_ECN_WPA_PSK = 0x02,
   WIFI_ECN_WPA2_PSK = 0x03,
   WIFI_ECN_WPA_WPA2_PSK = 0x04,
*/
#ifndef WIFI_SECURITY_TYPE
//#error "Symbol WIFI_SECURITY_TYPE must be defined."
#define WIFI_SECURITY_TYPE WIFI_ECN_WPA2_PSK
#endif /* WIFI_SECURITY_TYPE  */

#define TERMINAL_USE

#ifndef RETRY_TIMES
#define RETRY_TIMES 3
#endif

#define WIFI_FAIL 1
#define WIFI_OK   0


//Prototypes
int wifi_setup(void);

#endif //__WIFISETUP_H
```

3.  Save the file.
4.  Under \AzureRTOS-NetXDuo\Core\Src, create a file called wifi_setup.c.
5.  Open wifi_setup.c and enter the following code:

```c
#include <string.h>
#include <stdlib.h>
#include <stdio.h>
#include <stdbool.h>
#include "stm32l4xx_hal.h"
#include "stm32l4s5i_iot01.h"
#include "wifi.h"
#include "main.h"
#include "wifi_setup.h"

#ifdef TERMINAL_USE
extern ES_WIFIObject_t      EsWifiObj;
#endif /* TERMINAL_USE */


extern  SPI_HandleTypeDef hspi;


typedef enum
{
  WS_IDLE = 0,
  WS_CONNECTED,
  WS_DISCONNECTED,
  WS_ERROR,
} WebServerState_t;

uint8_t  MAC_Addr[6];
uint8_t  IP_Addr[4];
uint8_t  Gateway_Addr[4];
uint8_t  DNS1_Addr[4];
uint8_t  DNS2_Addr[4];


int  wifi_setup(void)
{

  uint32_t  retry_connect=0;

  /* Enable execution profile.  */
  CoreDebug -> DEMCR |= CoreDebug_DEMCR_TRCENA_Msk;
  DWT -> CTRL |= DWT_CTRL_CYCCNTENA_Msk;


  /*Initialize and use WIFI module */
  if(WIFI_Init() ==  WIFI_STATUS_OK)
  {

#if defined (TERMINAL_USE)
      /* Lib info.  */
```

103

```c
    uint32_t hal_version = HAL_GetHalVersion();
    uint32_t bsp_version = BSP_GetVersion();

    printf("STM32L4XX Lib:\r\n");
    printf("> CMSIS Device Version: %d.%d.%d.%d.\r\n",
__STM32L4_CMSIS_VERSION_MAIN, __STM32L4_CMSIS_VERSION_SUB1,
__STM32L4_CMSIS_VERSION_SUB2, __STM32L4_CMSIS_VERSION_RC);
    printf("> HAL Driver Version: %ld.%ld.%ld.%ld.\r\n",
((hal_version >> 24U) & 0xFF), ((hal_version >> 16U) & 0xFF),
((hal_version >> 8U) & 0xFF), (hal_version & 0xFF));
    printf("> BSP Driver Version: %ld.%ld.%ld.%ld.\r\n",
((bsp_version >> 24U) & 0xFF), ((bsp_version >> 16U) & 0xFF),
((bsp_version >> 8U) & 0xFF), (bsp_version & 0xFF));

    /* ES-WIFI info.  */
    printf("ES-WIFI Firmware:\r\n");
    printf("> Product Name: %s\r\n", EsWifiObj.Product_Name);
    printf("> Product ID: %s\r\n", EsWifiObj.Product_ID);
    printf("> Firmware Version: %s\r\n", EsWifiObj.FW_Rev);
    printf("> API Version: %s\r\n", EsWifiObj.API_Rev);
#endif /* TERMINAL_USE */

    if(WIFI_GetMAC_Address(MAC_Addr) == WIFI_STATUS_OK)
    {
#if defined (TERMINAL_USE)
        printf("ES-WIFI MAC Address: %X:%X:%X:%X:%X:%X\r\n",
MAC_Addr[0], MAC_Addr[1], MAC_Addr[2], MAC_Addr[3], MAC_Addr[4],
MAC_Addr[5]);
#endif /* TERMINAL_USE */
    }
    else
    {
#if defined (TERMINAL_USE)
        printf("!!!ERROR: ES-WIFI Get MAC Address Failed.\r\n");
#endif /* TERMINAL_USE */
        return WIFI_FAIL;
    }

    while((retry_connect++) < RETRY_TIMES)
    {
        printf("wifi connect try %ld times\r\n",retry_connect);
        if( (WIFI_Connect(WIFI_SSID, WIFI_PASSWORD,
WIFI_SECURITY_TYPE) == WIFI_STATUS_OK))
    {
#if defined (TERMINAL_USE)
        printf("ES-WIFI Connected.\r\n");
#endif /* TERMINAL_USE */

        if(WIFI_GetIP_Address(IP_Addr) == WIFI_STATUS_OK)
```

```c
        {
#if defined (TERMINAL_USE)
        printf("> ES-WIFI IP Address: %d.%d.%d.%d\r\n",
IP_Addr[0], IP_Addr[1], IP_Addr[2], IP_Addr[3]);

        if(WIFI_GetGateway_Address(Gateway_Addr) ==
WIFI_STATUS_OK)
        {
            printf("> ES-WIFI Gateway Address: %d.%d.%d.%d\r\n",
Gateway_Addr[0], Gateway_Addr[1], Gateway_Addr[2],
Gateway_Addr[3]);
        }

        if(WIFI_GetDNS_Address(DNS1_Addr,DNS2_Addr) ==
WIFI_STATUS_OK)
        {
            printf("> ES-WIFI DNS1 Address: %d.%d.%d.%d\r\n",
DNS1_Addr[0], DNS1_Addr[1], DNS1_Addr[2], DNS1_Addr[3]);

            printf("> ES-WIFI DNS2 Address: %d.%d.%d.%d\r\n",
DNS2_Addr[0], DNS2_Addr[1], DNS2_Addr[2], DNS2_Addr[3]);
        }
#endif /* TERMINAL_USE */

        if((IP_Addr[0] == 0)&& (IP_Addr[1] == 0)&& (IP_Addr[2]
== 0)&& (IP_Addr[3] == 0)){
            return WIFI_FAIL;
        }


    }
    else
    {
#if defined (TERMINAL_USE)
        printf("!!!ERROR: ES-WIFI Get IP Address Failed.\r\n");
#endif /* TERMINAL_USE */
        return WIFI_FAIL;
    }

        break;
    }
  }
    if(retry_connect > RETRY_TIMES)
    {

#if defined (TERMINAL_USE)
        printf("!!!ERROR: ES-WIFI NOT connected.\r\n");
#endif /* TERMINAL_USE */
        return WIFI_FAIL;
    }
```

```
  }
  else
  {
#if defined (TERMINAL_USE)
    printf("!!!ERROR: ES-WIFI Initialize Failed.\r\n");
#endif /* TERMINAL_USE */
    return WIFI_FAIL;
  }


  return WIFI_OK;
}

__weak void user_button_callback()
{
}

void EXTI1_IRQHandler(void)
{
  HAL_GPIO_EXTI_IRQHandler(GPIO_PIN_1);
}

void SPI3_IRQHandler(void)
{
  HAL_SPI_IRQHandler(&hspi);
}

void HAL_GPIO_EXTI_Callback(uint16_t GPIO_Pin)
{
  switch (GPIO_Pin)
  {
    case (GPIO_PIN_1):
    {
      SPI_WIFI_ISR();
      break;
    }
    case (USER_BUTTON_PIN):
    {
      user_button_callback();
      break;
    }
    default:
    {
      break;
    }
  }
}
```

6.  Save the file.

The source code for both include and source files was taken directly from board_setup.c and modified to remove excess code. Make sure that you add your Wi-FI SSID and password.

### 9.8.2 Initialize NetX Duo

The sample_netx_duo_ping.c contains the code to initialize NetX Duo. The call to initialize is made in the tx_application_define() function. We will create a thread that calls the NetX Duo initialization.

1. In the AzureRTOS-NetXDuo project, open app_threadx.c.
2. Add the following includes at line 25in the "USER CODE BEGIN Includes" section:

```
#include "app_azure_rtos.h"
#include "nx_api.h"
#include "wifi.h"
#include "nxd_dns.h"
#include "nx_secure_tls_api.h"
#include "stm3214xx_hal.h"
#include <stdio.h>
#include "main.h"
#include "nx_driver_stm3214.h"
```

3. Add the type definitions at about line 37 in the "USER CODE BEGIN PTD" section:

```
NX_PACKET_POOL      pool_0;
NX_IP               ip_0;


#ifndef SAMPLE_PACKET_COUNT
#define SAMPLE_PACKET_COUNT             (20)
#endif /* SAMPLE_PACKET_COUNT */

#ifndef SAMPLE_PACKET_SIZE
#define SAMPLE_PACKET_SIZE             (1200)  /* Set the
default value to 1200 since WIFI payload size
(ES_WIFI_PAYLOAD_SIZE) is 1200.  */
#endif /* SAMPLE_PACKET_SIZE  */

#define SAMPLE_POOL_SIZE                ((SAMPLE_PACKET_SIZE +
sizeof(NX_PACKET)) * SAMPLE_PACKET_COUNT)
#define SAMPLE_IP_STACK_SIZE           (2048)


/* Define the stack/cache for ThreadX.  */
static UCHAR sample_pool_stack[SAMPLE_POOL_SIZE];
```

```
static      ULONG      sample_ip_stack[SAMPLE_IP_STACK_SIZE      /
sizeof(ULONG)];
```

4. Add the following to set the thread stack size at line 61 in the private define section:

```
/* USER CODE BEGIN PD */
#define THREAD_STACK_SIZE 1024
/* USER CODE END PD */
```

5. Add the private variables and the function prototype for the thread at line 72:

```
/* Private variables ---------------------------------------------
--------------*/
/* USER CODE BEGIN PV */
uint8_t thread_stack[THREAD_STACK_SIZE];
TX_THREAD thread_ptr;
/* USER CODE END PV */

/* Private function prototypes ----------------------------------
--------------*/
/* USER CODE BEGIN PFP */
VOID nx_connect_thread_entry(ULONG intial_input);
/* USER CODE END PFP */
```

6. In App_ThreadX_Init we need to add the call to create the thread in the "USER CODE BEGIN App_ThreadX_Init" section after the pool variable declaration:

```
/* USER CODE BEGIN App_ThreadX_Init */
(void)byte_pool;
tx_thread_create(&thread_ptr, "nx_connect_thread",
nx_connect_thread_entry, 0x1234, thread_stack,
THREAD_STACK_SIZE, 15,15,1,TX_AUTO_START);
/* USER CODE END App_ThreadX_Init */
```

7. Finally, in the "USER CODE BEGIN 1" section add the code for the thread function:

```
VOID nx_connect_thread_entry(ULONG initial_input){
    printf("nx connect thread entry reached\n");
    printf("\n");

  UINT   status;
  /* Initialize the NetX system.   */
  nx_system_initialize();

  /* Create a packet pool.   */
```

```
    status = nx_packet_pool_create(&pool_0, "NetX Main Packet
Pool", SAMPLE_PACKET_SIZE, sample_pool_stack ,
SAMPLE_POOL_SIZE);

    /* Check for pool creation error.  */
    if (status)
    {
        printf("PACKET POOL CREATE FAIL.\r\n");
        return;
    }

    /* Create an IP instance.  */
    status = nx_ip_create(&ip_0, "NetX IP Instance 0", 0, 0,
&pool_0, nx_driver_stm3214, (UCHAR*)sample_ip_stack,
sizeof(sample_ip_stack), 1);

    /* Check for IP create errors.  */
    if (status)
    {
        printf("IP CREATE FAIL.\r\n");
        return;
    }
}
```

8. Save the file.

**Note**: Since we are only going to ping the network connection, creating the packet pool and ip instances are not really needed. Ping uses ICMP to send messages. Initializing NetX Duo is done for completeness and leads to the next project.


### 9.8.3   Add Call to initialize Wi-Fi
We need to call the wifi_setup() function in main.c before the call to start the kernel.

1. Open main.c.
2. Add the function prototype to the "USER CODE BEGIN PFP" section around line 94:

```
/* USER CODE BEGIN PFP */
void wifi_setup(void);
/* USER CODE END PFP */
```

3. Add the cal to the wifi_setup() function in the USER CODE BEGIN 2 section around line 149:

```
/* USER CODE BEGIN 2 */
wifi_setup();
/* USER CODE END 2 */
```

4.  Save the file.

### 9.8.4   Define printf() and scanf() calls to go out and in the debug port

1.  Open STM32CubeIDE.
2.  The first code we want to add is to configure the printf() function to send data out of the debug port. In the AzureRTOS-NetXDuo project, open main.h and add the putchar_prototype declaration code around line 228:

```
/* USER CODE BEGIN Private defines */
#define PUTCHAR_PROTOTYPE int __io_putchar(int ch)
#define GETCHAR_PROTOTYPE int __io_getchar(void)
/* USER CODE END Private defines */
```

3.  Save and close main.h.
4.  Open main.c and add the putchar_prototype code at about line 1009

```
/* USER CODE BEGIN 4 */
PUTCHAR_PROTOTYPE
{
  /* Place your implementation of fputc here */
  /* e.g. write a character to the USART1 and Loop until the end
of transmission */
  HAL_UART_Transmit(&huart1, (uint8_t *)&ch, 1, 0xFFFF);

  return ch;
}

GETCHAR_PROTOTYPE
{
      uint8_t ch;
      HAL_UART_Receive(&huart1, &ch, 1, HAL_MAX_DELAY);

      /* Echo character back to console */
      HAL_UART_Transmit(&huart1, &ch, 1, HAL_MAX_DELAY);

      /* And cope with Windows */
      if (ch == '\r') {
          uint8_t ret = '\n';
          HAL_UART_Transmit(&huart1, &ret, 1, HAL_MAX_DELAY);
      }
```

```
    return ch;
}
/* USER CODE END 4 */
```

### 9.8.5 Add Random Number Generator Code

The random number generator will be used for security in the project. The implementation of the random number generator for NetX Duo in the Example is a little off. There are two implementations. One is working and one is unused.

1. In main.c after the PUTCHAR_PROTOYPE code add the following:

```
int hardware_rand(void)
{
    HAL_StatusTypeDef status;
    uint32_t rand = 0;
    status = HAL_RNG_GenerateRandomNumber(&hrng, &rand);

    if(status){
      printf("Random Generator status: %i\n", status);
    }

    return rand;
}
```

2. At line number 96 after the wifi_setup() function prototype declaration, add the prototype definition for the random number generator function.

```
/* USER CODE BEGIN PFP */
void wifi_setup(void);
int hardware_rand(void);
/* USER CODE END PFP */
```

3. We can test the call to the function on startup. After the wifi_setup() function is called, print out a random number and a message that board setup is completed:

```
/* USER CODE BEGIN 2 */
wifi_setup();
printf("Random number: %u\n", hardware_rand());
printf("\n");
printf("********************************\n");
printf("Board initialization completed\n");
printf("********************************\n");
printf("\n");
/* USER CODE END 2 */
```

4. At the top of the file in the private includes section, add stdio.h:

```
/* Private includes ---------------------------------------------
---------------*/
/* USER CODE BEGIN Includes */
#include <stdio.h>
/* USER CODE END Includes */
```

5. Save the file.
6. Open nx_port.h found in Middlewares\ST\netxduo\ports\cortex_m4\gnu\inc.
7. Set the #define for NX_RAND to the random number function

```
extern int hardware_rand(void);
#define NX_RAND             hardware_rand
```

8. Save the file.


## 9.9   Build and Debug

We are now ready to build and test the project.

1. Right-click on the AzureRTOS-NetXDuo.
2. Select "Build Project" from the context menu. The project should build without error.
3. Connect the STM32L4S5 Discovery Kit to the development computer using the USB cable.
4. Start the debug session. The debugger will deploy the .elf file to the board and reboot the board. The debugger will stop at the HAL_Init() function.
5. Start ABCOMTerm or a similar serial terminal program and connect to the board's COM port.
6. In STM32CubeIDE, click on the continue button ▷ to let the program run. The terminal program will show the output from the STM32L4S5 Discovery Kit.

STM32L4XX Lib:
> CMSIS Device Version: 1.7.2.0.
> HAL Driver Version: 1.13.3.0.
> BSP Driver Version: 1.0.0.0.
ES-WIFI Firmware:
> Product Name: Inventek eS-WiFi
> Product ID: ISM43362-M3G-L44-SPI
> Firmware Version: C3.5.2.5.STM
> API Version: v3.5.2
ES-WIFI MAC Address: C4:7F:51:91:44:40
wifi connect try 1 times
ES-WIFI Connected.
> ES-WIFI IP Address: 192.168.1.41
> ES-WIFI Gateway Address: 192.168.1.1

> ES-WIFI DNS1 Address: 192.168.1.1
> ES-WIFI DNS2 Address: 8.8.8.8
Random number: 1989723014


\*\*\*\*\*\*\*\*\*\*\*\*\*\*\*\*\*\*\*\*\*\*\*\*\*\*\*\*\*\*\*\*\*\*
Board initialization completed
\*\*\*\*\*\*\*\*\*\*\*\*\*\*\*\*\*\*\*\*\*\*\*\*\*\*\*\*\*\*\*\*\*\*


nx connect thread entry reached

7. With the successful WiFi connection, open a command prompt and ping the IP address of the STM32L4S5 Discovery Kit and you should get a reply.

C:\Users\Tester_>ping 192.168.1.41

Pinging 192.168.1.41 with 32 bytes of data:
Reply from 192.168.1.41: bytes=32 time=209ms TTL=255
Reply from 192.168.1.41: bytes=32 time=3ms TTL=255
Reply from 192.168.1.41: bytes=32 time=3ms TTL=255
Reply from 192.168.1.41: bytes=32 time=3ms TTL=255

Ping statistics for 192.168.1.41:
    Packets: Sent = 4, Received = 4, Lost = 0 (0% loss),
Approximate round trip times in milli-seconds:
    Minimum = 3ms, Maximum = 209ms, Average = 54ms

8. Stop the debugger when finished.

## 9.10 Summary: One Step Closer to the Cloud

Creating an ThreadX (Azure RTOS) NetX Duo project in STM32Cube is a little more involved, but there is useful source code available to create a project. From this project, we are getting closer to addressing the driving question for this book. ThreadX components will be constantly upgraded, which will not be translated immediately into components in the STM32Cube repository; but we can use the STM32Cube tools to define the HAL layer and pull in the latest ThreadX components from GitHub. Now, we are ready to create a project that connects to Azure IoT Central.

# 10 Project 6: Azure IoT Central Connection

The first 5 projects have walked through the basics of getting a project set up and exploring the different features and available source code to include in a project. The previous project shows how complex it can be to integrate the Azure IoT C SDK and a new version of the NetX Duo component into a project. We have a solution that connects to the network and gets an IP address. For this project, we will build on the last project to connect to Azure and send data to Azure IoT Central Application.

## 10.1 Review: The Two-Sample Azure RTOS Projects.

In chapter 3, you download two sample example projects. Both examples come with different solutions to connect to Azure.

The Getting Started Example supported different MCUs and different boards. The end result was sensor data being sent to an Azure IoT Central application. Several articles on Annabooks.com demonstrated how to set up the projects in Visual Studio Code to build and debug the application. The article "Azure RTOS and STMicroelectronics STM32 Discovery Kit IoT (STM32L4S5)" specifically covers the STM32 Discovery Kit IoT. The whole project structure is a mess with support files everywhere to accommodate all the different MCUs. There are wrapper files for Azure IoT C SDK to make calls simpler in the application, and some of the code is not used at all. After all the code setup and connections, only one file contains all the work performed by the firmware: nx_client.c. In the end, the resulting firmware does connect to the Azure IoT Central application and uses a template for the STM32 Discovery Kit IoT to present the data visually. Trying to extract the exact setup in the STM32Cube tools is impossible.

The STM32CubeIDE Example is more straightforward. We were able to repurpose the Ping project in the last chapter rather easily. There are two other projects in the example: sample_azure_iot_embedded_sdk and sample_azure_iot_embedded_sdk_pnp. The sample_azure_iot_embedded_sdk project sends a raw data message to an Azure IoT HUB connection. It is possible to connect to Azure IoT Central via Azure IoT HUB, but a virtual network and endpoint are needed to complete the connection. The sample_azure_iot_embedded_sdk_pnp project is set up for device provisioning services and can connect to Azure IoT Central, but it sends fake temperature data.

As the title of the project implies, we will create a project to connect to Azure IoT Central. We will be drawing the source code from the STM32CubeIDE

115

sample_azure_iot_embedded_sdk_pnp example. We will add the sensor drivers so real data can be sent to the cloud.


## 10.2 Initiate the Project with STM32CubeMX

Now, we are ready to create the project. The project will use the ThreadX and NetX Duo software packages in the STM32Cube repository.

1. Open STM32CubeMX.
2. Under the New Project, click on "Access to Board Selector".
3. A new window appears. From the Commercial Part Number drop-down, select B-L4S5I-IOT01A. If you are using a different STM32 board, then select that product number.
4. On the right-hand side, there is a list of boards associated with the part number. There is only one, click on the item. The pane above files shows the information about the board. From here you can click on the links to get more resources about the board and STM32L4+ MCU on the board.
5. Click on "Start Project" in the top right corner.
6. You will be asked to initialize the default settings. Click Yes. The project gets initialized and a picture of the MCU appears with all the pins and associated I/O configuration for the development board already laid out. Since this is a development board, we will keep the defaults.
7. Now, we need to add Azure RTOS to the project. Click on Software Packs->Select Components

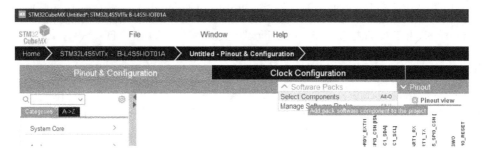

8. The package selector appears. There are some packages active and some that are inactive. The active packages are for the STM32 MCU on the board. The inactive packages are for other STM32 MCUs. Locate STMicroelectronics.X-CUBE-AZRTOS-L4 and click the Install button next to the package. This will install the Azure RTOS package support for the STM32 MCU that is on the board.
9. Once installed, expand the branches under RTOS ThreadX->ThreadX.
10. Tick the box next to Core.
11. Expand the branches under Network NetXDuo->NetXDuo.
12. Tick the boxes for NX Core, DNS, MQTT, Cloud, SNTP, WebClient, TLS, and Crypto.

| | | | |
|---|---|---|---|
| ∨ STMicroelectronics X-CUBE-AZRTOS-L4 | ⊘ | 2.0.0 ∨ | |
| ∨ *RTOS* ThreadX | ⊘ | 6.2.0 | |
| ∨ ThreadX | ⊘ | | |
| Core | ⊘ | 6.2.0 | ☑ |
| PerformanceInfo | | 6.2.0 | ☐ |
| TraceX support | | 6.2.0 | ☐ |
| Low Power support | | 6.2.0 | ☐ |
| > *File System* FileX | | 6.2.0 | |
| > *File System* LevelX | | 6.2.0 | |
| > *File System* Interfaces | | 2.1.0 | |
| > *USB* USBX | | 6.2.0 | |
| ∨ *Network* NetXDuo | ⊘ | 6.2.0 | |
| ∨ NetXDuo | ⊘ | | |
| NX Core | ⊘ | 6.2.0 | ☑ |
| Addons AutoIP | | 6.2.0 | ☐ |
| Addons DHCP Client | | 6.2.0 | ☐ |
| Addons DHCP Server | | 6.2.0 | ☐ |
| Addons DNS | ⊘ | 6.2.0 | ☑ |
| Addons mDNS | | 6.2.0 | ☐ |
| Addons MQTT | ⊘ | 6.2.0 | ☑ |
| Addons NAT | | 6.2.0 | ☐ |
| Addons Cloud | ⊘ | 6.2.0 | ☑ |
| Addons PPP | | 6.2.0 | ☐ |
| Addons SNTP | ⊘ | 6.2.0 | ☑ |
| Addons Web Client | ⊘ | 6.2.0 | ☑ |
| Addons Web Server | | 6.2.0 | ☐ |
| TLS | ⊘ | 6.2.0 | ☑ |
| Crypto | ⊘ | 6.2.0 | ☑ |
| > *Network* Interfaces | | 2.1.0 | |

13. Click the OK button in the bottom right corner.
14. STMicroelectronics.X-CUBE-AZRTOS-L4 is now listed under Middleware and Software Packs. Click on STMicroelectronics.X-CUBE-AZRTOS-L4.
15. The STMicroelectronics.X-CUBE-AZRTOS-L4 Mode and Configuration will appear. Check the boxes for both the RTOS ThreadX and Network NetXDuo. The configuration options are displayed below the Mode. We will keep the defaults.

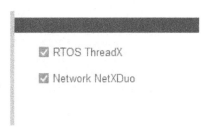

16. We need to configure the clock. Expand the System Core on the left side.
17. The SYS-tick is used by the HAL and ThreadX. To separate the two, we will give the HAL a different clock source. In the categories, select System Core->SYS.
18. In the Timebase Source drop-down, select TIM6.

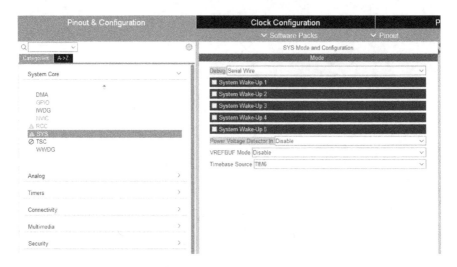

19. Finally, we need to add a HAL feature to help with security. In the Categories tab, expand the Security section and enable RNG (Random Number Generator).

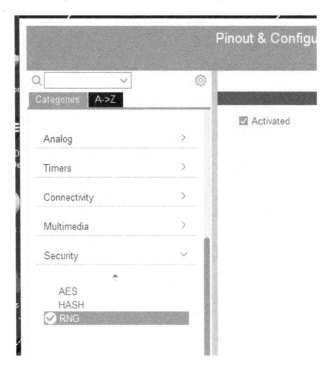

20. Click on the Project Manager tab.

21. Fill in the following:
    a. Project Name: AzureRTOS-IoT-Central.
    b. Project Location: Make the project location the same location as the STM32CubeIDE Workspace folder.
    c. Toolchain/IDE: STM32CubeIDE.
    d. Click on the "Generate Under Root" checkbox.

22. Click on "Generate Code" in the top right. The project will be created in the folder you selected.
23. Once the code has been generated, a dialog will ask you to open the files or open the project. Click on the "Open Project" button.
24. STM32CubeIDE will open and import the project into the workspace folder that you created in Chapter 3.
25. Close STM32CubeMX.

## 10.3 Add Sensor and WiFi Driver Source Code

Project 4 covered the sensor related to barometer: Temperature, Pressure, and Humidity. We will copy these driver source files over to the project, and then add a thread that will read the sensors as a test.

1. Open File Explorer, and copy the sensor source file in \Barometer-NoRTOS\Drivers\BSP\B-L4S5I-IOT01 folder to the \AzureRTOS-IoT-Central\Drivers\BSP\B-L4S5I-IOT01 folder.

```
Name
  stm32l4s5i_iot01.c
  stm32l4s5i_iot01.h
  stm32l4s5i_iot01_hsensor.c
  stm32l4s5i_iot01_hsensor.h
  stm32l4s5i_iot01_psensor.c
  stm32l4s5i_iot01_psensor.h
  stm32l4s5i_iot01_tsensor.c
  stm32l4s5i_iot01_tsensor.h
```

2. Copy the \Barometer-NoRTOS\Drivers\BSP\Components folder to the \AzureRTOS-IoT-Central\Drivers\BSP folder.
3. Copy the \AzureRTOS-NetXDuo\Drivers\BSP\WiFi folder to the \AzureRTOS-IoT-Central\Drivers\BSP\components folder

4. In STM32CubeIDE, right-click on AzureRTOS-IoT-Central project and select Refresh from the context menu. You should see the newly added files.

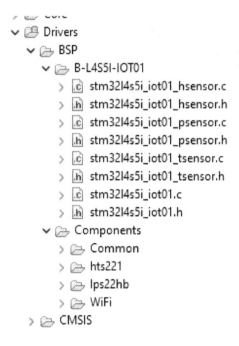

5. In Project Explorer, right-click on each of the folders under AzureRTOS-IoT-Central\Drivers\BSP\Components folder, select "add/remove include path…", and click the OK button to include the paths for both debug and release builds.
6. Open main.c, if it is not already open; and add the following includes for the sensors:

```
/* USER CODE END Header */
/* Includes ------------------------------------------------------
---------------*/
#include "app threadx.h"
#include "main.h"
#include "stm3214s5i_iot01.h"
#include "stm3214s5i_iot01_hsensor.h"
#include "stm3214s5i_iot01_psensor.h"
#include "stm3214s5i_iot01_tsensor.h"
```

7.  Now, we are going to add a function to initialize the sensors. First, add the Sensors_init prototype after the wifi_setup function prototype:

```
/* USER CODE BEGIN PFP */
void wifi_setup(void);
static void Sensors_init(void);
/* USER CODE END PFP */
```

8.  After the putchar and getchar functions, add the code for the Sensor_init at around line 1043.

```
static void Sensors_init(void)
{

    if (HSENSOR_OK != BSP_HSENSOR_Init())
    {
        printf("ERROR: BSP_HSENSOR_Init\r\n");
    }

    if (TSENSOR_OK != BSP_TSENSOR_Init())
    {
        printf("ERROR: BSP_TSENSOR_Init\r\n");
    }

    if (PSENSOR_OK != BSP_PSENSOR_Init())
    {
        printf("ERROR: BSP_PSENSOR_Init\r\n");
    }
}
```

9.  In the main() function, add the call to initialize the sensors and the Green LED after the call to wifi_setup() and before the call toe MX_ThreadX_Init().

121

```
/* USER CODE BEGIN 2 */
wifi_setup();

//Initialize the Green LEd
BSP_LED_Init(LED_GREEN);

//Initialize the sensors
Sensors_init();

/* USER CODE END 2 */


MX_ThreadX_Init();
```

10. Save the file.
11. Open app_threadx.c
12. Since we are going to be reading from the sensors, we need to add their includes.

```
#include <stdio.h>
#include "stm3214s5i_iot01.h"
#include "stm3214s5i_iot01_hsensor.h"
#include "stm3214s5i_iot01_psensor.h"
#include "stm3214s5i_iot01_tsensor.h"
```

13. We will now add a thread that will be called once to print out the readings from the sensors. In the private variables section, add the variables to help configure the thread after the variables for the first azure thread.

```
/* Private variables ---------------------------------------------
---------------*/
/* USER CODE BEGIN PV */
uint8_t thread_stack[THREAD_STACK_SIZE];
TX_THREAD thread_ptr;

ULONG sensor_thread_stack[THREAD_STACK_SIZE / sizeof(ULONG)];
TX_THREAD sensor_thread;
```

14. Add the sensor_thread_entry prototype:

```
/* USER CODE BEGIN PFP */
VOID nx_connect_thread_entry(ULONG intial_input);
VOID sensor_thread_entry(ULONG intial_input);
/* USER CODE END PFP *
```

15. Add the sensor thread creation at line 99, after the creation of the first thread.

```
/* USER CODE BEGIN App_ThreadX_Init */
(void)byte_pool;
tx_thread_create(&thread_ptr, "nx_connect_thread",
nx_connect_thread_entry, 0x1234, thread_stack,
THREAD_STACK_SIZE, 15,15,1,TX_AUTO_START);
tx_thread_create(&sensor_thread,"Sensor
Thread",sensor_thread_entry,0x8085,          sensor_thread_stack,
THREAD_STACK_SIZE, 15, 15, 1,TX_AUTO_START);
```

16. Add the sensor_thread_entry function after the nx_connect_thread_entry function:

```
void sensor_thread_entry(ULONG paramter){

    printf("Reading Sensors\n");
    printf("Temperature: %f\n", BSP_TSENSOR_ReadTemp());
    printf("Humidity: %f\n", BSP_HSENSOR_ReadHumidity());
    printf("Pressure: %f\n",  BSP_PSENSOR_ReadPressure());


}
```

17. Save the file.

## 10.4 Add Missing NetX Duo Addons and Separate the Sample

In Chapter 3, you downloaded the latest NetX Duo component from Git Hub. We need to add two missing items to the project.

1. Open File Explorer.
2. Go to the directory that has netxduo0-master.
3. In the Git\netxduo-master\addons folder, copy azure_iot and websocket folders and pass them to the \AzureRTOS-IoT-Central\Middlewares\ST\netxduo\addons

azure_iot

cloud

dns

mqtt

sntp

web

websocket

4. In the azure_iot folder, there is a samples folder. Copy and paste the folder to the \Git folder.
5. In STM32CubeIDE, right-click on the AzureRTOS-IoT-Central and select Refresh. The two add-ons will be available.
6. Now, we need to remove some unnecessary items. The Azure IOT SDK for C has a number of sample and test projects included. If these are left in place, the build will fail. In File Explorer, go to AzureRTOS-NetXDuo\Middlewares \ST\netxduo\addons\azure_iot\azure-sdk-for-c\sdk.
7. Delete the "samples" and "tests" folders.
8. Next, delete the \AzureRTOS-NetXDuo2\Middlewares\ST\netxduo\addons\azure_iot \azure-sdk-for-c\sdk\src\azure\platform folder.
9. From the menu, select Project->Properties.
10. A Properties dialog appears. Expand the C/C++ Build and click on Settings.
11. Go to MGU GCC Compiler->Include Paths.

You will see a number of include paths already set up for the HAL, threadx, and NetX Duo. Now, we need to add all the paths for NetX Duo. You could add each folder individually through this interface or you could simply right-click on the folder and select Add/Remote include path from the context menu.

12. To make things a little easier, in the Exercise in Chapter 9 that comes with the book, there is an IncludePaths.txt file. Open the file in Notepad or Notepad++. The file contains all of the includes:

**../Core/Inc**
**../AZURE_RTOS/App**
**../NetXDuo/App**
**../Drivers/STM32L4xx_HAL_Driver/Inc**
**../Drivers/STM32L4xx_HAL_Driver/Inc/Legacy**
**../Drivers/CMSIS/Device/ST/STM32L4xx/Include**

```
../Drivers/CMSIS/Include
../Middlewares/ST/netxduo/addons/mqtt/
../Middlewares/ST/netxduo/addons/dns/
../Middlewares/ST/netxduo/addons/web/
../Middlewares/ST/netxduo/common/inc/
../Middlewares/ST/netxduo/ports/cortex_m4/gnu/inc/
../Middlewares/ST/netxduo/nx_secure/inc/
../Middlewares/ST/netxduo/nx_secure/ports/
../Middlewares/ST/netxduo/crypto_libraries/inc/
../Middlewares/ST/netxduo/crypto_libraries/ports/cortex_m4/gnu/inc/
../Middlewares/ST/netxduo/addons/cloud/
../Middlewares/ST/threadx/common/inc/
../Middlewares/ST/threadx/ports/cortex_m4/gnu/inc/
../Middlewares/ST/netxduo/addons/sntp/
"${workspace_loc:/${ProjName}/Middlewares/ST/netxduo/addons/azure_iot/azur
e_iot_security_module/iot-security-module-core/deps/flatcc/include}"
"${workspace_loc:/${ProjName}/Middlewares/ST/netxduo/addons/azure_iot/azur
e_iot_security_module/iot-security-module-core/deps/flatcc/src/runtime}"
"${workspace_loc:/${ProjName}/Middlewares/ST/netxduo/addons/azure_iot/azur
e-sdk-for-c/sdk/inc}"
"${workspace_loc:/${ProjName}/Middlewares/ST/netxduo/addons/azure_iot}"
"${workspace_loc:/${ProjName}/Middlewares/ST/netxduo/ports/cortex_m4/gnu/i
nc}"
"${workspace_loc:/${ProjName}/Middlewares/ST/netxduo/nx_secure/inc}"
"${workspace_loc:/${ProjName}/Middlewares/ST/netxduo/nx_secure/ports}"
"${workspace_loc:/${ProjName}/Middlewares/ST/netxduo/crypto_libraries/inc}"
"${workspace_loc:/${ProjName}/Middlewares/ST/netxduo/common/inc}"
"${workspace_loc:/${ProjName}/Middlewares/ST/netxduo/addons/dns}"
"${workspace_loc:/${ProjName}/Middlewares/ST/netxduo/addons/mqtt}"
"${workspace_loc:/${ProjName}/Middlewares/ST/netxduo/addons/web}"
"${workspace_loc:/${ProjName}/Middlewares/ST/netxduo/addons/cloud}"
"${workspace_loc:/${ProjName}/Middlewares/ST/netxduo/addons/azure_iot/azur
e_iot_security_module}"
"${workspace_loc:/${ProjName}/Middlewares/ST/netxduo/addons/azure_iot/azur
e_iot_security_module/inc}"
"${workspace_loc:/${ProjName}/Middlewares/ST/netxduo/addons/azure_iot/azur
e_iot_security_module/inc/configs/RTOS_BASE}"
"${workspace_loc:/${ProjName}/Middlewares/ST/netxduo/addons/azure_iot/azur
e_iot_security_module/iot-security-module-core/inc}"
"${workspace_loc:/${ProjName}/Middlewares/ST/netxduo/addons/websocket}"
"${workspace_loc:/${ProjName}/Middlewares/ST/netxduo/addons/azure_iot}"
"${workspace_loc:/${ProjName}/Drivers/BSP/B-L4S5I-IOT01}"
"${workspace_loc:/${ProjName}/Drivers/BSP/Components/Common}"
"${workspace_loc:/${ProjName}/Middlewares/ST/netxduo/addons/sntp}"
"${workspace_loc:/${ProjName}/Drivers/BSP/Components/hts221}"
"${workspace_loc:/${ProjName}/Drivers/BSP/Components/lps22hb}"
"${workspace_loc:/${ProjName}/Drivers/BSP/Components/WiFi}"
```

13. Copy all the contents, and past the contents to the Include paths section by hitting a CTRL+V on the keyboard.

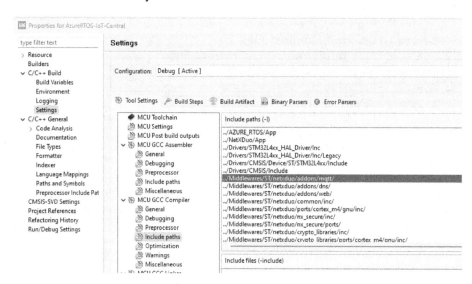

14. Click on the "Apply and Close" button.

## 10.5 Edit the Code for Main.c

The main.c contains all the board setup code, so we will add the debug port, random number generator, and sensor setup as we have done in the last two chapters.

1. In STM32CubeIDE, open main.h and add the GETCHAR_PROTOTYPE after the PUTCHAR_PROTOTYPE:

```
/* USER CODE BEGIN Private defines */
#define PUTCHAR_PROTOTYPE int __io_putchar(int ch)
#define GETCHAR_PROTOTYPE int __io_getchar(void)
/* USER CODE END Private defines */
```

2. Save and close the file.
3. Open main.c.
4. Add the following private includes:

```
/* Private includes ------------------------------------
---------------*/
/* USER CODE BEGIN Includes */
#include "stm3214s5i_iot01.h"
#include "stm3214s5i_iot01_tsensor.h"
#include "stm3214s5i_iot01_hsensor.h"
#include "stm3214s5i_iot01_psensor.h"
```

126

```
#include "wifi_setup.h"
#include <stdio.h>
#include <math.h>
#include <stdlib.h>
/* USER CODE END Includes */
```

5. Add the putchar, getchar, random number generator, and sensor initialization functions to "USER CODE BEGIN 4" section:

```
/* USER CODE BEGIN 4 */

PUTCHAR_PROTOTYPE
{
  /* Place your implementation of fputc here */
  /* e.g. write a character to the USART1 and Loop until the end
of transmission */
  HAL_UART_Transmit(&huart1, (uint8_t *)&ch, 1, 0xFFFF);

  return ch;
}

GETCHAR_PROTOTYPE
{
    uint8_t ch;
    HAL_UART_Receive(&huart1, &ch, 1, HAL_MAX_DELAY);

    /* Echo character back to console */
    HAL_UART_Transmit(&huart1, &ch, 1, HAL_MAX_DELAY);

    /* And cope with Windows */
    if (ch == '\r') {
        uint8_t ret = '\n';
        HAL_UART_Transmit(&huart1, &ret, 1, HAL_MAX_DELAY);
    }

    return ch;
}


int hardware_rand(void)
{
    HAL_StatusTypeDef status;
    uint32_t rand = 0;
    status = HAL_RNG_GenerateRandomNumber(&hrng, &rand);

    if(status){
      printf("Random Generator status: %i\n", status);
    }
```

```
        return rand;
}


static void Sensors_Init(void)
{

    if (HSENSOR_OK != BSP_HSENSOR_Init())
    {
        printf("ERROR: BSP_HSENSOR_Init\r\n");
    }

    if (TSENSOR_OK != BSP_TSENSOR_Init())
    {
        printf("ERROR: BSP_TSENSOR_Init\r\n");
    }

    if (PSENSOR_OK != BSP_PSENSOR_Init())
    {
        printf("ERROR: BSP_PSENSOR_Init\r\n");
    }
}

/* USER CODE END 4 */
```

6. Scroll back up and add the prototypes for the random number and sensor initialization functions:

```
/* USER CODE BEGIN PFP */
static void Sensors_Init(void);
int hardware_rand(void);
/* USER CODE END PFP */
```

7. Finally, let's add the code to initialize the sensors and WiFi. Set up the random number generator, and output the current sensor readings in the USER CODE BEGIN 2 section:

```
/* USER CODE BEGIN 2 */
//Initialize the sensors
Sensors_Init();
//Enable the WiFi
wifi_setup();

printf("\n");
//SEED the random number generator library
srand(hardware_rand());
printf("RNG test %u\n", hardware_rand());
printf("Random Number Library Test %u\n", rand());
```

128

```
HAL_GPIO_TogglePin(LED2_GPIO_Port, LED2_Pin);
HAL_Delay(1000);

printf("\nTest sensor readings\n");
//Get Temperature
float temp_value = roundf(BSP_TSENSOR_ReadTemp()*100)/100;
printf("Temperature (degC): %0.2f\n", temp_value);
HAL_Delay(1000);
float pres_value = roundf(BSP_PSENSOR_ReadPressure()*100)/100;
printf("Pressure (hPa): %0.2f\n", pres_value);
HAL_Delay(1000);
float humid_value = roundf(BSP_HSENSOR_ReadHumidity()*100)/100;
printf("Humidity (%%rH): %0.2f\n", humid_value);
printf("\n");
printf("********************************\n");
printf("Board initialization completed\n");
printf("********************************\n");
printf("\n");

/* USER CODE END 2 */
```

8. Save the file.

## 10.6 Add Defines to nx_port.h

There are some conditional compiler options that need to be defined or the build will fail.

1. In STM32CubeIDE, open nx_port.h found under \AzureRTOS-IoT-Central\Middlewares\ST\netxduo\ports\cortex_m4\gnu\inc.
2. Add the following defines:

```
#define NX_ENABLE_IP_PACKET_FILTER 1
#define NXD_MQTT_CLOUD_ENABLE 1
#define NX_ENABLE_EXTENDED_NOTIFY_SUPPORT 1
#define NX_ENABLE_TCPIP_OFFLOAD 1
#define NX_ENABLE_INTERFACE_CAPABILITY 1
#define NX_SECURE_ENABLE 1
```

3. Save the file.

**Warning**: If you have to go back to STM32CubeMX to make changes, the changes in nx_port.h will be deleted.

## 10.7 Add the Azure IoT Sample Source Code from the NetX Duo Component Download

Attempting to pull code from either of the example projects is a tedious and fruitless exercise. Creating an application from scratch using the Azure IoT C SDK is doable, but takes a little time. Also, connecting, sending, receiving, and constantly monitoring the connection to Azure takes some code work. It just so happens that Microsoft provides the application source code to perform all these tasks as part of the NetX Duo component download. We pulled out the samples directory when we added the components to the azure_iot addon in section 10.4. The samples folder contains the equivalent code that is found in the STM32CubeIDE Examples. The sample_azure_iot_embedded_sdk.c and sample_azure_iot_embedded_sdk_pnp.c are the same two project samples from the STM32CubeIDE Examples. The source code in this folder has all the working code to connect, send, receive, and monitor a connection to either Azure IoT Hub or Azure IoT Central.

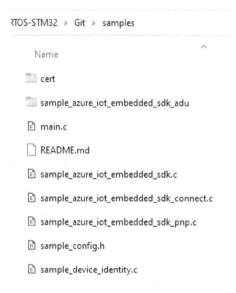

The main.c file has the code to call board_setup.c and then start the kernel. Looking closer, this file is actually, the app_threadx.c and AZURE_RTOS app all wrapped into one. The code can be modified so that the main.c generated by STM32CubeMX performs the call to start the kernel in this file. The next question is which project file do we use: sample_azure_iot_embedded_sdk.c or sample_azure_iot_embedded_sdk_pnp.c? The sample_azure_iot_embedded_sdk_pnp.c is more suited to connecting to Azure IoT Central and sending sensor data, so we will use that file. The sample_azure_iot_embedded_sdk_adu folder contains the source code for the Azure Device Update solution, we will not use this solution for this project. The cert folder contains the generic certs that we will use in the project. The high-level diagram for our project looks as follows.

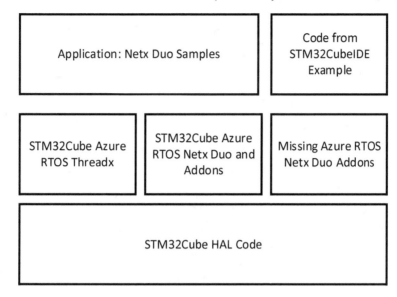

1. In File Explorer, go to the \Git\samples folder.
2. Make a backup copy of main.c (i.e. main.c.bak).
3. Rename main.c to app_threadx.c.
4. From the \Hit\samples folder and cert subfolder, copy nx_azure_iot_cert.h, nx_azure_iot_ciphersuites.h, and sample_config.h to \AzureRTOS-IoT-Central\Core\Inc folder.
5. From the \Hit\samples folder and cert subfolder, copy app_threadx.c, nx_azure_iot_cert.c, nx_azure_iot_ciphersuites.c, sample_azure_iot_embedded_sdk_connect.c, sample_azure_iot_embedded_sdk_pnp.c, and sample_device_identity.c to the AzureRTOS-IoT-Central\Core\Src folder.
6. In STMCubeIDE, right-click on the project and select Refresh.

```
✓ 🗂 Core
  ✓ 🗁 Inc
    > 🗎 app_threadx.h
    > 🗎 main.h
    > 🗎 nx_azure_iot_cert.h
    > 🗎 nx_azure_iot_ciphersuites.h
    > 🗎 RTE_Components.h
    > 🗎 sample_config.h
    > 🗎 stm32l4xx_hal_conf.h
    > 🗎 stm32l4xx_it.h
    > 🗎 tx_user.h
    > 🗎 wifi_setup.h
  ✓ 🗁 Src
    > 🗎 app_threadx.c
    > 🗎 main.c
    > 🗎 nx_azure_iot_cert.c
    > 🗎 nx_azure_iot_ciphersuites.c
    > 🗎 sample_azure_iot_embedded_sdk_connect.c
    > 🗎 sample_azure_iot_embedded_sdk_pnp.c
    > 🗎 sample_device_identity.c
    > 🗎 stm32l4xx_hal_msp.c
    > 🗎 stm32l4xx_hal_timebase_tim.c
    > 🗎 stm32l4xx_it.c
    > 🗎 syscalls.c
    > 🗎 sysmem.c
    > 🗎 system_stm32l4xx.c
    > 🗎 tx_initialize_low_level.S
    > 🗎 wifi_setup.c
      🗎 app_threadx.c.bak
```

4.  Open app_threadx.c and add two defines at the top of the file:

```
#define SAMPLE_DHCP_DISABLE 1
#define SAMPLE_NETWORK_CONFIGURE sample_network_configure
```

5.  Rename void main() to MX_ThreadX_Init(). The contents of the function are to remain unchanged. Notice that the call to set up the board is greyed out as the main.c file performs the board setup. We can now see that the other goal for the book is achievable, so you can repeat these steps to support a different STM32 MCU.

```
/* Define main entry point.  */
void MX_ThreadX_Init()
{

#ifdef SAMPLE_BOARD_SETUP
    SAMPLE_BOARD_SETUP();
#endif /* SAMPLE_BOARD_SETUP */

    /* Enter the ThreadX kernel.  */
    tx_kernel_enter();
}
```

6. The last thing to do is pull in code from the STM32CubeIDE Example. Specifically, the sample_network.c file, found in \b-l4s5i-iot01a_2022_11_30\stm32cubeide \sample_azure_iot_embedded_sdk_pnp folder, needs to be pulled out and placed at the end of app_threadx.c to address the define we added a few steps ago. We could have just set the define to NX_SUCCESS and just skipped the code, but the original STM32CubeIDE has a good check in place.

```
VOID sample_network_configure(NX_IP *ip_ptr, ULONG *dns_address)
{
ULONG    ip_address = 0;
ULONG    network_mask = 0;
ULONG    gateway_address = 0;
UINT     status;
ULONG    dns_address_second;

    WIFI_GetIP_Address((UCHAR *)&ip_address);
    WIFI_GetIP_Mask((UCHAR *)&network_mask);
    WIFI_GetGateway_Address((UCHAR *)&gateway_address);
    NX_CHANGE_ULONG_ENDIAN(ip_address);
    NX_CHANGE_ULONG_ENDIAN(network_mask);
    NX_CHANGE_ULONG_ENDIAN(gateway_address);

    if (dns_address)
    {
        WIFI_GetDNS_Address((UCHAR *)dns_address, (UCHAR
*)&dns_address_second);
        NX_CHANGE_ULONG_ENDIAN(*dns_address);
    }

    status = nx_ip_address_set(ip_ptr, ip_address,
network_mask);

    /* Check for IP address set errors.  */
    if (status)
    {
        printf("IP ADDRESS SET FAIL.\r\n");
        return;
    }

    status = nx_ip_gateway_address_set(ip_ptr, gateway_address);

    /* Check for gateway address set errors.  */
    if (status)
    {
        printf("IP GATEWAY ADDRESS SET FAIL.\r\n");
        return;
    }
}
```

7. Save the file.
8. Open sample_azure_iot_embedded_sdk_pnp.c
9. Add the following includes to the file:

```
#include "stm3214s5i_iot01.h"
#include "stm3214s5i_iot01_tsensor.h"
#include "stm3214s5i_iot01_hsensor.h"
#include "stm3214s5i_iot01_psensor.h"
```

10. Scroll down until you see the Telemetry section. Around line 114. Add the following code for pressure and humidity after the temperature constant:

```
/* Telemetry.  */
static const CHAR telemetry_name[] = "temperature";
static const CHAR telemetry_pressure[] = "pressure";
static const CHAR telemetry_humidity[] = "humidity";
```

11. Scroll down to the sample_telemetry_action function. Modify the code to send the real sensor data for temperature, pressure, and humidity around line 814.

```
if (nx_azure_iot_json_writer_append_begin_object(&json_writer)
||
nx_azure_iot_json_writer_append_property_with_double_value(&json
_writer, (UCHAR *)telemetry_name, sizeof(telemetry_name) - 1,
BSP_TSENSOR_ReadTemp(), DOUBLE_DECIMAL_PLACE_DIGITS) ||
nx_azure_iot_json_writer_append_property_with_double_value(&json
_writer, (UCHAR *)telemetry_pressure, sizeof(telemetry_pressure)
- 1, BSP_PSENSOR_ReadPressure(), DOUBLE_DECIMAL_PLACE_DIGITS) ||
nx_azure_iot_json_writer_append_property_with_double_value(&json
_writer, (UCHAR *)telemetry_humidity, sizeof(telemetry_humidity)
- 1, BSP_HSENSOR_ReadHumidity(), DOUBLE_DECIMAL_PLACE_DIGITS) ||
nx_azure_iot_json_writer_append_end_object(&json_writer))
{
    printf("Telemetry message failed to build message\r\n");

nx_azure_iot_hub_client_telemetry_message_delete(packet_ptr);
    return;
}
```

12. Save the file.
13. Since the source code we added contains the threadx functions, we need to delete the threadx files that contain the same functions generated by STM32CubeMX. Under the AzureRTOS-IoT-Central project, right-click on AZURE_RTOS and select Delete from the context menu.

## 10.8  Create Azure IoT Central Application

Now, we need to set up the application on Azure IoT Central. Device connection and template model information will be collected and placed into the azure_config.h file.

1. In a browser, open https://apps.azureiotcentral.com/home

Note: If you do not have an Azure account, you will have to sign up for one. https://azure.microsoft.com and set up a subscription plan.

2. Sign into the account or create an account.
3. Click on Build App.
4. In the Custom app tile, click Create app.

   Application Name: STM32Barometer.
   Pricing Plan: <your choice> each plan lets you have 2 free devices for a limited number of messages.
   Billing Info: <your choice>.

5. Click Create.
6. Click "Add a device" and set the following:

   Device name: STM32Disco1
   Device ID: stm32disco1

7. Click on the Create button.
8. Now, we need to gather some information for the firmware. Click on Devices.
9. Click on STM32Disco1.
10. Click on the Connect button at the top of the page.
11. The ID scope, device id, and Primary key are shown in a dialog box. Copy each of these items to Notepad.

With the application set up, we can use the copied information in the firmware and test the firmware.

**Note**: Normally, we would create a template to associate with the device, but the sample application will use a sample template that is already available in Azure IoT Central. Look in sample_azure_iot_embedded_sdk_pnp.c, there is a define for the model ID.

```
#define SAMPLE_PNP_MODEL_ID "dtmi:com:example:Thermostat;4"
#define          SAMPLE_PNP_DPS_PAYLOAD          {\"modelId\":\""
SAMPLE_PNP_MODEL_ID "\"}"
```

Sample_config.h also has a setting for the MODULE_ID that is not used. You can modify the second define, use the MODULE_ID instead, and define your own template. You will need to get the interface @Id for the template, which can be found when you click on Edit Identity while editing the template. For this project, we will use the template; but make additions to support the new sensors.

## 10.9 Build and Debug

We need to add the connection information first, and then we can build and debug.

1. Open STM32CubeIDE.
2. Under the AzureRTOS-IoT-Central project, open the azure_config.h file.
3. Make sure that you have set up the WiFi SSID, password, and security type.
4. Fill in the other #defines with the information from the Azure IoT Central application.

| Constant name | Value |
|---|---|
| IOT_DPS_ID_SCOPE | ID scope value |
| IOT_DPS_REGISTRATION_ID | Device ID value |
| IOT_DEVICE_SAS_KEY | Primary key value |

5. Save the file.
6. Right-click on the AzureRTOS-NetXDuo.
7. Select "Build Project" from the context menu. The project should build without error.
8. Connect the STM32L4S5 Discovery Kit to the development computer using the USB cable.
9. Start the debug session. The debugger will deploy the .elf file to the board and reboot the board. The debugger will stop at the HAL_Init() function.
10. Start ABCOMTerm or a similar serial terminal program and connect to the board's COM port.

11. In STM32CubeIDE, click on the continue button ▷ to let the program run. You will see the application's output in the terminal:

STM32L4XX Lib:
> CMSIS Device Version: 1.7.2.0.
> HAL Driver Version: 1.13.3.0.
> BSP Driver Version: 1.0.1.0.
ES-WIFI Firmware:
> Product Name: Inventek eS-WiFi
> Product ID: ISM43362-M3G-L44-SPI
> Firmware Version: C3.5.2.5.STM
> API Version: v3.5.2
ES-WIFI MAC Address: C4:7F:51:91:44:40
wifi connect try 1 times
ES-WIFI Connected.
> ES-WIFI IP Address: 192.168.1.41

> ES-WIFI Gateway Address: 192.168.1.1
> ES-WIFI DNS1 Address: 192.168.1.1
> ES-WIFI DNS2 Address: 8.8.8.8

RNG test 2644040502

Random Number Library Test 740743301

Test sensor readings
Temperature (degC): 26.01
Pressure (hPa): 1001.03
Humidity (%rH): 47.53


\*\*\*\*\*\*\*\*\*\*\*\*\*\*\*\*\*\*\*\*\*\*\*\*\*\*\*\*\*\*\*\*\*\*
Board initialization completed
\*\*\*\*\*\*\*\*\*\*\*\*\*\*\*\*\*\*\*\*\*\*\*\*\*\*\*\*\*\*\*\*\*\*

IP address: 192.168.1.41
Mask: 255.255.255.0
Gateway: 192.168.1.1
DNS Server address: 192.168.1.1
SNTP Time Sync...0.pool.ntp.org
SNTP Time Sync successfully.
[INFO] Azure IoT Security Module has been enabled, status=0
Start Provisioning Client...
[INFO] IoTProvisioning client connect pending
Registered Device Successfully.
IoTHub Host Name: iotc-cd9e424b-9ec3-4d72-a2e3-49f37ee4ecfc.azure-devices.net; Device ID: stm32disco1.
Connected to IoTHub.
Sent properties request.
Telemetry message send: {"temperature":26.77,"pressure":1001.03,"humidity":46.57}.
[INFO] Azure IoT Security Module message is empty
Received all properties
Telemetry message send: {"temperature":26.85,"pressure":1001.05,"humidity":46.46}.

12. Telemetry messages are being sent to the Azure IoT Central application. In the Azure IoT Central STM32Baromater application, open the STM32Disco1 device. You will see the raw data coming in.

138

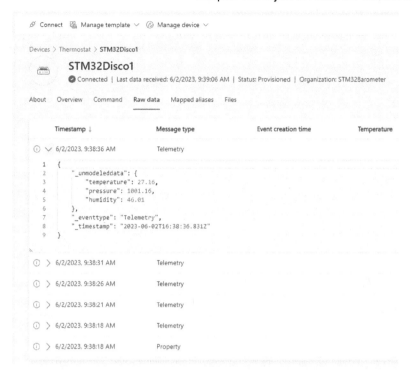

13. Click on Overview, and you will see the temperature history being displayed.

14. To add the pressure and humidity information, click on Manage template->Edit Template. You will see how the default Thermostat template is configured.

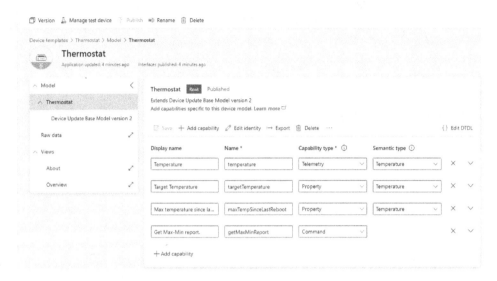

15. Click +Add capability and fill in the following information for pressure:

    Display Name: **Pressure**
    Name: **pressure**
    Semantic type: **Pressure**
    Unit: **Kilopascal**
    Display unit: **kPa**
    Description: **Pressure in kPa**

16. Click +Add capability and fill in the following information for humidity:

    Display Name: **Humidity**
    Name: **humidity**
    Semantic type: **Humidity**
    Unit: **Percent**
    Display unit: **%RH**
    Description: **Humidity in %RH**

17. Click Save. This will save the new parameters.
18. Under Model, click on Overview so we can change the visual aspects.
19. We can add the two new parameters to the big diagram for Temperature. Click on the Pencil symbol.
20. A side panel appears.
21. Change the title to Temperature, Pressure, Humidity.
22. Add the capability for Pressure and Humidity.

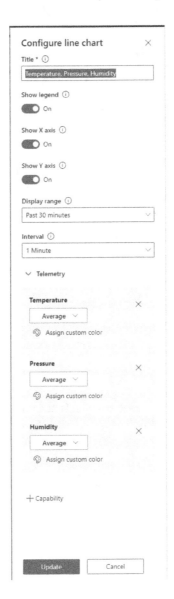

23. Click Update.

24. In the tile, click on the resize icon ▣ and set the size to 3 : 3.
25. For the individual temperature box, click on the three dots (...).
26. Select Change Visualization to Last Known Value.
27. Click on the three dots (...) and set Size Available to 1:1.
28. Under Edit view, click on "Start with devices".
29. Under Telemetry, select Pressure and click "Add tile".

30. A new tile appears for Pressure, change the tile to Last Known Value and resize to 1:1.
31. Repeat steps 28-30 for Humidity.
32. The tile can be re-arranged

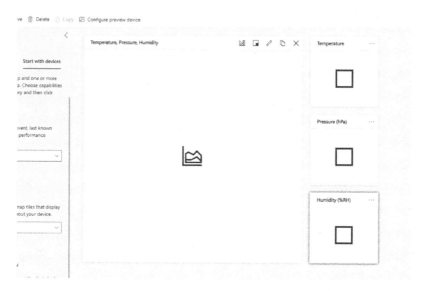

33. Click Save.
34. Click Back.
35. Click Publish and a dialog appears with a summary of changes. Click Publish.
36. Go back to Devices->STM32Disco1. The Overview has been updated to include the two new sensors.

37. Hit the reset button on the board so we can restart sending data.

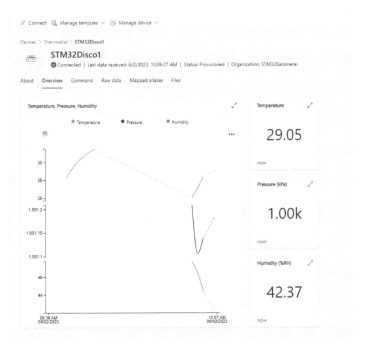

## *10.10 Summary: Question Answered*

We have finally addressed one of the book's driving questions. The STM32Cube repository doesn't have all the features for NetX Duo, but the extra add-ons can be added to a project from the NetX Duo component git download. The sample application provided in the NetX Duo component is a good starting point for getting connected to the Azure IoT Central application. The next step would be to add gyroscopic accelerometer sensors and a command to control the Green LED to the project and the Azure IoT Central application. Creating a custom Azure IoT Central template would be beneficial, rather than relying on the stock template. Now that we have addressed the first guiding question, let's address the second guiding question and get Azure RTOS running on a different STM32 MCU Development boards.

# 11 Project 7: FileX

FileX is the next ThreadX component to explore. This project will be broken down into two parts using two different boards. The first part will use the Nucleo-H723ZG board that features the STM32H723ZG MCU, which is a Cortex-M7 processor. Unlike the Discovery board we have used so far, Nucleo platforms are barebones boards with the ST-Link connection, MCU, and header connectors for the I/O from the MCU. Some Nucleo boards, like the Nucleo-H723ZG, have Ethernet support. The reason for going with this board is to replicate an STMicroelectronics webinar class that demonstrates using FileX with SRAM as the storage disk. There will be some modifications to the webinar information to make the project more aligned with the STM32CubeIDE project structure. By using the RAM disk, you can visually see data in RAM, which is great for learning, but most OEMs will want to store data between reboots. Therefore, the second part of this project is to move to the STM32U5 Discovery Kit (B-U585I-IOT02A) and implement the same FileX application using the onboard flash chip for storage rather than RAM. As you will see, there are some little things to address that the RAM drive project doesn't cover.

## 11.1 Using RAM, Create Project Part 1 with STM32CubeMX

Using STM32CubeMX, we will create a project for the Nucleo-H723ZG:

1. Open STM32CubeMX.
2. Under the New Project, click on "Access to Board Selector".
3. A new window appears. From the Commercial Part Number drop-down, select Nucleo-H723ZG. If you are using a different STM32 board, then select that product number.
4. On the right-hand side, there is a list of boards associated with the part number. There is only one, click on that item.

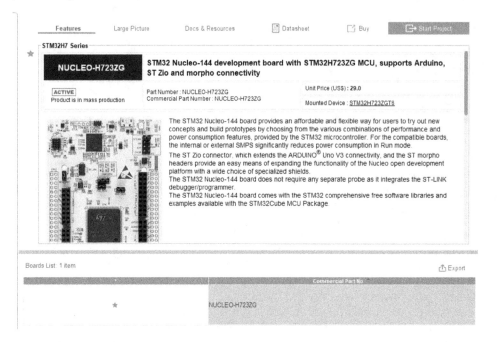

5. Click on "Start Project" in the top right corner.
6. You will be asked to initialize the default settings. Click Yes. The project gets initialized and a picture of the MCU appears with all the pins and associated I/O configuration for the development board already laid out. Since this is a development board, we will keep the defaults.
7. Now, we need to add Azure RTOS ThreadX and FileX components to the project. Click on Software Packs->Select Components.
8. The package selector appears. There are some packages active and some that are inactive. The active packages are for the STM32 MCU on the board. The inactive packages are for other STM32 MCUs. Locate STMicroelectronics.X-CUBE-AZRTOS-H7 and click the Install button next to the package. This will install the Azure RTOS package support for the STM32 MCU that is on the board.
9. Once installed, expand the branches under RTOS ThreadX->ThreadX.
10. Tick the box next to Core.
11. Expand File System- > FileX.
12. Tick the box next to FileX/Core.
13. Expand File System -> Interfaces.
14. Tick the box next to FileX Internal RAM interface.

146

| | | | |
|---|---|---|---|
| > STMicroelectronics.X-CUBE-AZRTOS-G0 | | 1.1.0 | Install |
| > STMicroelectronics.X-CUBE-AZRTOS-G4 | | 2.0.0 ∨ | Install |
| ∨ STMicroelectronics.X-CUBE-AZRTOS-H7 | ⊘ | 3.1.0 ∨ | |
| ∨ RTOS ThreadX | ⊘ | 6.2.0 | |
| ∨ ThreadX | ⊘ | | |
| Core | ⊘ | 6.2.0 | ☑ |
| PerformanceInfo | | 6.2.0 | ☐ |
| TraceX support | | 6.2.0 | ☐ |
| Low Power support | | 6.2.0 | ☐ |
| ∨ File System FileX | ⊘ | 6.2.0 | |
| FileX / Core | ⊘ | 6.2.0 | ☑ |
| > File System LevelX | | 6.2.0 | |
| ∨ File System Interfaces | ⊘ | 2.1.0 | |
| FileX Internal RAM interface | ⊘ | 2.1.0 | ☑ |
| FileX SD interface | | 2.1.0 | ☐ |
| FileX MMC interface | | 2.1.0 | ☐ |
| FileX LevelX NOR interface | | 2.1.0 | ☐ |
| FileX LevelX NAND interface | | 2.1.0 | ☐ |
| FileX Custom interface | | 2.1.0 | ☐ |

15. Click the OK button in the bottom right corner.
16. STMicroelectronics.X-CUBE-AZRTOS-H7 is now listed under Middleware and Software Packs. Click on STMicroelectronics.X-CUBE-AZRTOS-H7.
17. The STMicroelectronics.X-CUBE-AZRTOS-H7 Mode and Configuration will appear. Tick all the boxes for RTOS ThreadX, FileSystem FileX, and File System Interfaces.

18. The configuration options are displayed below the Mode. We will keep the defaults, and we will make a couple of changes. The first is to change the memory pool size. In FileX memory pool size, change the value from 1024 to 4096. If you click on the "I" in the top right corner, the information for the parameter is listed below. There is a range that the value can be set.

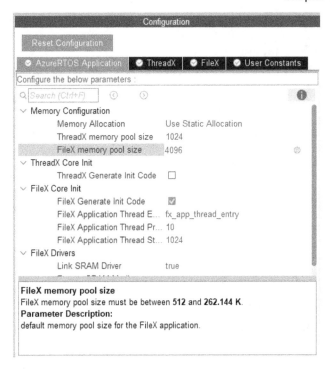

19. Now, we need to change the starting address of the FileX RAM drive; so it doesn't conflict with RAM that is already used. Click on the FileX tab.
20. Scroll down and change the SRAM Disk Address to D1_AXISRAM2_BASE. Notice that the size of the SRAM disk is 8192 KB.

21. We need to configure the clock. Expand the System Core on the left side.
22. The SYS-tick is used by the HAL and Azure RTOS. To separate the two, we will give the HAL a different clock source. In the categories, select System Core->SYS.
23. In the Timebase Source drop-down, select TIM6.

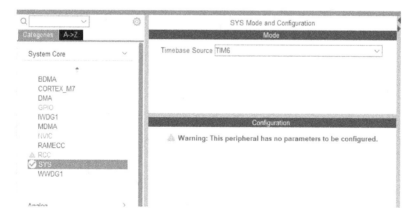

24. Click on the Project Manager tab.
25. Fill in the following:

      a. Project Name: AzureRTOS-FileX-H7.
      b. Project Location: Make the project location the same location as the STM32CubeIDE Workspace folder.
      c. Toolchain/IDE: STM32CubeIDE.
      d. Click on the "Generate Under Root" checkbox.

26. Click on "Generate Code" in the top right. The project will be created in the folder you selected.
27. Once the code has been generated, a dialog will ask you to open the files or open the project. Click on the "Open Project" button.
28. STM32CubeIDE will open and import the project into the workspace folder that you created in Chapter 3.
29. Close STM32CubeMX.

## 11.2 FileX Additions to the Project

Expand the newly created project within STMCubeIDE, so we can go over the FileX-specific file additions.

The Azure RTOS FileX component files are placed in the Middlewares\ST\filex folder. Like ThreadX and NetX Duo, FileX also creates an application folder, which contains a thread definition that will be called on system reset. The main function calls the MX_ThreadX_Init function in app_threadx.c, which calls the tx_kernel_enter function in tx_initialize_kernel_enter.c. The tx_kernel_enter calls the tx_application_define, which not only calls the App_ThreadX_Init function in app_threadx.c, but also calls the MX_FileX_Init function that is in app_filex.c, which kicks off the main FileX thread.

We made a couple of changes in STM32CubeMX for FileX. The FileX memory pool size can be found in the app_azure_rtos_config.h file. What about the base address setting? The FX_SRAM_DISK_BASE_ADDRESS setting was changed to D1_AXISRAM2_BASE. The D1_AXISRAM2_BASE definition value can be found in stm32h723xx.h. The value is set to 0x24020000UL. The FX_SRAM_DISK_BASE_ADDRESS is defined in the fx_stm32_sram_driver.h. For completeness, look in the linker script file. RAM_D1 is the base memory address which is set to 0x24000000. You wouldn't have known about the D1_AXISRAM2_BASE definition until you created the project, opened the stm32h723xx.h file, and went back to make the change.

## 11.3 *Writing the Application.*

The online seminar video does a sleight of hand in copying over an app_filex.c file to replace what was generated by STM32CubeMX. The actual code does work, but it is not aligned with the code that is generated from the STM32CubeMX tool. Since we want to repeat the operation for other platforms, we are going to use what is generated to fill in the application with the necessary application code, which we will reuse with a flash disk in part 2.

1. Sending messages out the console port is helpful to track the operation of the FileX thread. In STM32CubeIDE, open main.h and add the prototype for PUTCHAR and GETCHAR to the "USER CODE BEGIN Private defines" section:

```
/* USER CODE BEGIN Private defines */
#define PUTCHAR_PROTOTYPE int __io_putchar(int ch)
#define GETCHAR_PROTOTYPE int __io_getchar(void)
/* USER CODE END Private defines */
```

2. Save and close the file.
3. Open main.c and add the putchar and getchar functions to the "USER CODE BEGIN 4" section:

```
/* USER CODE BEGIN 4 */
PUTCHAR_PROTOTYPE
{
/* Place your implementation of fputc here */
/* e.g. write a character to the USART1 and Loop until the end
of transmission */
HAL_UART_Transmit(&huart3, (uint8_t *)&ch, 1, 0xFFFF);

return ch;
}

GETCHAR_PROTOTYPE
{
    uint8_t ch;
    HAL_UART_Receive(&huart3, &ch, 1, HAL_MAX_DELAY);
```

```
    /* Echo character back to console */
    HAL_UART_Transmit(&huart3, &ch, 1, HAL_MAX_DELAY);

    /* And cope with Windows */
    if (ch == '\r') {
        uint8_t ret = '\n';
        HAL_UART_Transmit(&huart3, &ret, 1, HAL_MAX_DELAY);
    }

    return ch;
}
/* USER CODE END 4 */
```

4. Save and close the file.
5. Open app_fileX.c file.
6. In the "USER CODE BEGIN Includes" section add the following:

```
/* USER CODE BEGIN Includes */
#include "main.h"
#include <stdio.h>
/* USER CODE END Includes */
```

7. In the "USER CODE BEGIN PV" section, define a FX_FILE type:

```
/* USER CODE BEGIN PV */
FX_FILE                     fx_file;
/* USER CODE END PV */
```

8. In the "USER CODE BEGIN 0" section add the following:

```
/* USER CODE BEGIN 0 */
printf("FX Initialization...\n");
/* USER CODE END 0 */
```

9. Now we move down to the main FileX thread. In the "USER CODE BEGIN fx_app_thread_entry 0" section, add the following:

```
/* USER CODE BEGIN fx_app_thread_entry 0 */
ULONG bytes_read;
ULONG available_space_post;
CHAR read_buffer[32];
CHAR data[] = "Hello from FileX on STM32\n";

printf("Staring FileX App Thread.\n");
/* USER CODE END fx_app_thread_entry 0 */
```

The code defines some variables that will be used to read the file, a string to send to the file, and a variable to get the available space on the disk.

153

10. Finally, we will add the actual application that creates a file in the RAM disk, opens the file, writes to the file, and closes the file. Then the application then re-opens the file, reads the file, prints the message to the console, and closes the file. The final step is to get some information about the disk and close the RAM disk. In the "USER CODE BEGIN fx_app_thread_entry 1", add the following:

```
/* USER CODE BEGIN fx_app_thread_entry 1 */

///////Create and open the file///////////
sram_status = fx_file_create(&sram_disk, "FXTEST.TXT");
if (sram_status != FX_SUCCESS)
{
        if (sram_status != FX_ALREADY_CREATED)
          {
            /* Create error, call error handler.   */
            Error_Handler();
          }
}

sram_status =  fx_file_open(&sram_disk, &fx_file, "FXTEST.TXT",
FX_OPEN_FOR_WRITE);
if (sram_status != FX_SUCCESS)
{
   Error_Handler();
}
////move the file pointer to the start of the file////////
sram_status =  fx_file_seek(&fx_file, 0);
if (sram_status != FX_SUCCESS)
{
   Error_Handler();
}
/////////Write the message to teh file and close the file/////
sram_status =  fx_file_write(&fx_file, data, sizeof(data));
if (sram_status != FX_SUCCESS)
{
   Error_Handler();
}

sram_status =  fx_file_close(&fx_file);
if (sram_status != FX_SUCCESS)
{

   Error_Handler();
}

/////Open the File, print the message to the console, and close
the file////////
```

```
sram_status =  fx_file_open(&sram_disk, &fx_file, "FXTEST.TXT",
FX_OPEN_FOR_READ);
if (sram_status != FX_SUCCESS)
{
   Error_Handler();
}


sram_status =  fx_file_seek(&fx_file, 0);
if (sram_status != FX_SUCCESS)
{
   Error_Handler();
}


sram_status =  fx_file_read(&fx_file, read_buffer, sizeof(data),
&bytes_read);
if ((sram_status != FX_SUCCESS) || (bytes_read != sizeof(data)))
{
   Error_Handler();
}

printf("Here is the file contents:  %s", read_buffer);


sram_status =  fx_file_close(&fx_file);
if (sram_status != FX_SUCCESS)
{
   Error_Handler();
}


////Gets some stats on the disk and close the disk////////
sram_status =  fx_media_space_available(&sram_disk,
&available_space_post);
if (sram_status != FX_SUCCESS)
{
   Error_Handler();
}

printf("Space available: %lu\n", available_space_post);

sram_status =  fx_media_close(&sram_disk);
if (sram_status != FX_SUCCESS)
{
   Error_Handler();
}

printf("Leaving FileX App Thread.\n");

/* USER CODE END fx_app_thread_entry 1 */
```

11. Save and close the file.

## 11.4 Debug the Applications on the Board

We are ready to test the application on the board.

1.  Build the application and make sure there are no errors.
2.  In the app_filex.c file, set a breakpoint at the line where the file is created:

```
181    ///////Create and open the file////////////
182    sram_status = fx_file_create(&sram_disk, "FXTEST.TXT");
183    if (sram_status != FX_SUCCESS)
```

3.  Make sure the Nucleo-H723ZG is connected to the development machine via the USB cable and start the debug session.
4.  The debugger will stop at HAL_Init().
5.  Open an Annabooks COM Terminal or equivalent serial terminal application.
6.  Set the COM port to the Nucleo-H723ZG USB COM port and initiate a connection.
7.  Hit the debug continue button. The code will run until the breakpoint is hit.
8.  Hit he debug step-over function. The program will run until the file is closed after writing.
9.  We want to see the contents of the file in memory. We want to add the memory monitor. From the menu, select Window->Show View->Memory.
10. In the memory monitor, click on the plus (+) sign and enter the starting address to the ram disk: 0x24020000

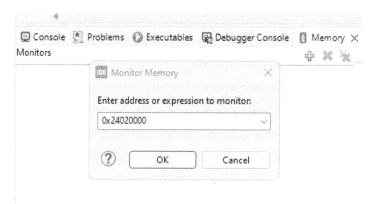

11. Click OK.
12. We want to see the data, in the memory monitor, click on the + New Renderings, and select traditional.

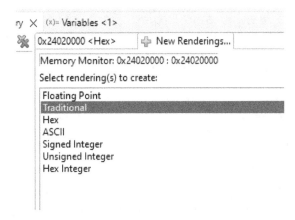

13. Click the "Add Renderings" button on the right. The traditional tab is created starting at 0x24020000. The sectors for the disk are 512 or 0x200 in size. At memory address 0x24020000, you will see the start of the RAM disk.

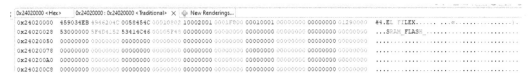

14. Scroll down to 0x24020400 and you will see the file name:

15. Scroll down to 0x24020800 and you will see the contents of the file:

16. Click continue to let the code run to the end. In the terminal you will see the full output from the printf statements:

FX Initialization...
Staring FileX App Thread.
Here is the file contents:  Hello from FileX on STM32
Space available: 0
Leaving FileX App Thread.

17. Stop the debugger.

## 11.5 Using NOR Flash, Create Project Part 2 with STM32CubeMX

Part 1 has demonstrated how to create a project with FileX. By recreating the sample from the STMicroelectronics webinar, we know the FileX application works. The next step is to implement persistent storage in the FileX project.

### 11.5.1 Why not use the STM32L4S5 Discovery Kit?

The next logical step would have been to repeat the same solution to access the MX25R6435F NOR-flash chip on the STM32L4S5 Discovery Kit. Unfortunately, the STM32Cube tools have an issue. The MX25R6435F is connected via QuadSPI, but there is no QuadSPI HAL support available even though there is a stm32l4xx_hal_qspi.c file available in the repository. At the time of this writing, there is no easy way to enable FileX to access the MX25R6435F via the tools. You could attempt to manually add all the correct source files with source code, but let's use the tools for a board that works. Maybe the QuadSPI hal driver will get fixed in the future.

**Note**: During the editing of the book, an ST Community member did post a possible solution to the STM32L4S5 Discovery Kit QuadSPI problem. The solution has yet to be tested.

### 11.5.2 Moving forward with STM32U5 Discovery Kit

For part 2, we will use the STM32U5 Discovery Kit (B-U585I-IOT02A) with the onboard MX25LM51245G NOR flash. Both kits are similar in concept, but the STM32U5 Discovery Kit is newer and has the latest support as of this writing.

1. Open STM32CubeMX.
2. Under the New Project, click on "Access to Board Selector".
3. A new window appears. From the Commercial Part Number drop-down, B-U585I-IOT02A.
4. On the right-hand side, there is a list of boards associated with the part number. There is only one, click on that item.

5. Click on "Start Project" in the top right corner.
6. You will be asked to initialize the default settings. Click Yes. The project gets initialized and a picture of the MCU appears with all the pins and associated I/O configuration for the development board already laid out. Since this is a development board, we will keep the defaults.
7. You will then be asked to turn on TrustZones. Leave the default as "without TrustZone Activated"., Click OK.

Now, we need to add ThreadX and FileX components to the project. Unlike the previous kits, the ThreadX (Azure RTOS) components are listed in the categories->Middleware and Software Packages for the B-U585I-IOT02A Discovery Kit.

8.   Click on THREADX and tick the checkbox next to Core.

9.   Click on LEVELX and tick the box next to "LevelX NOR Flash Support".
10.  Expand "File System Interfaces" and tick the box next to "OctoSPI memory interface".

160

11. In the configuration below, set the OctoSPI Instance to OCTOSPI2. Looking at the board schematics the NOR flash is connected to OCTOSPI2 and the SRAM chip is connected to OCTOSPI1.

12. Set the "Drivers Calls OctoSPI after init" setting to true.

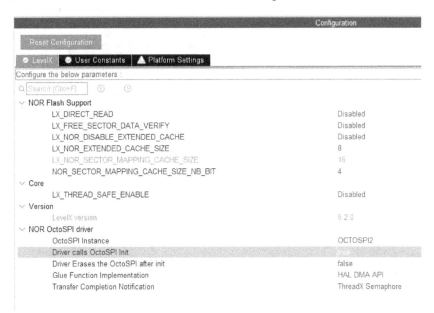

13. Click on the "Platform Settings" tab and set the OSPI Component to MX25LM51245G.

14. Click on FILEX and tick the box next to FileX Core.
15. Expand the "File System Interfaces" and select "LevelX NOR interface".

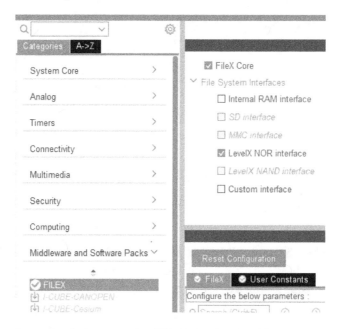

16. In the configuration below, set the "FileX Application Thread Stack Size" to 2048.
17. Set the "FileX memory pool size" to 4096.

**Note**: Increasing the size for both the application stack and pool size is important. Otherwise, the application will crash when formatting the media.

162

18. The drive is accessed via DMA, so we need to configure a DMA channel for data transfer. Expand the "System Core" category and click on GPDMA1.
19. There are 15 DMA channels with the last 4 using 8-word transfers. Set Channel 0 to "Standard Request Mode"
20. In the configuration below, click on the CH0 tab.
21. Set the Request Configuration->Request to OCTOSPI2.

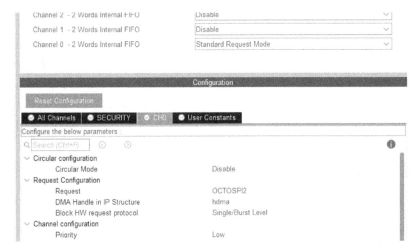

22. We need to configure the clock. Expand the System Core on the left side.
23. The SYS-tick is used by the HAL and Azure RTOS. To separate the two, we will give the HAL a different clock source. In the categories, select System Core->SYS.
24. In the Timebase Source drop-down, select TIM6.

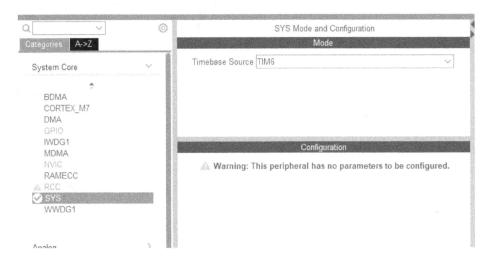

25. Click on NVIC.
26. Tick the box next to "OCTOSPI2 global interrupt".

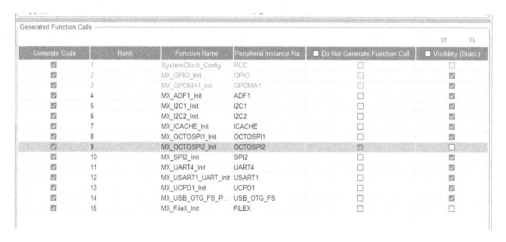

27. Click on the Project Manager tab.
28. Click on the Advanced settings on the left.
29. Under the "Generated Function Calls", check the box under the "Do Not Generate Function" column for MX_OCTOPI2_Init.
30. Uncheck the box under the Visibility (Static) column for MX_OCTOPI2_Init. The call into LEVELX will initiate the OCTOSPI2 port.

31. Click on the Project settings on the left. Fill in the following:

    e. Project Name: AzureRTOS-FileX-U585.
    f. Project Location: Make the project location is the same location for the STM32CubeIDE Workspace folder.
    g. Toolchain/IDE: STM32CubeIDE.
    h. Click on the "Generate Under Root" checkbox.

32. Click on "Generate Code" in the top right. The project will be created in the folder you selected.
    a. Once the code has been generated, a dialog will ask you to open the files or open the project. Click on the "Open Project" button.
33. STM32CubeIDE will open and import the project into the workspace folder that you created in Chapter 3.
34. Close STM32CubeMX.

## 11.6 FileX and LevelX Additions to the Project

The previous project created the file system in RAM. FileX performed direct calls to the RAM driver. Since flash devices can wear out over time, LevelX provides a wear-leveling function that extends the life of the flash. For a flash device, FileX makes calls to LevelX. LevelX makes calls down to the driver for the flash chip. If you expand the project, you can see the additional files that LevelX brings in.

## 11.7 *Writing the Application.*

We will reuse the application code in part 1 to write to the flash.

1. Sending messages out the console port is helpful to track the operation of the FileX thread. In STM32CubeIDE, open main.h and add the prototype for PUTCHAR and GETCHAR to the "USER CODE BEGIN Private defines" section:

```
/* USER CODE BEGIN Private defines */
#define PUTCHAR_PROTOTYPE int __io_putchar(int ch)
#define GETCHAR_PROTOTYPE int __io_getchar(void)
/* USER CODE END Private defines */
```

2. Save and close the file.
3. Open main.c and add the putchar and getchar functions to the "USER CODE BEGIN 4" section:

**Note**: Unlike Nucleo-H723ZG, the STM32U5 used huart1 for the COM port.

```
/* USER CODE BEGIN 4 */
PUTCHAR_PROTOTYPE
{
/* Place your implementation of fputc here */
/* e.g. write a character to the USART1 and Loop until the end
of transmission */
HAL_UART_Transmit(&huart1, (uint8_t *)&ch, 1, 0xFFFF);

return ch;
}

GETCHAR_PROTOTYPE
{
        uint8_t ch;
        HAL_UART_Receive(&huart1, &ch, 1, HAL_MAX_DELAY);

        /* Echo character back to console */
        HAL_UART_Transmit(&huart1, &ch, 1, HAL_MAX_DELAY);

        /* And cope with Windows */
        if (ch == '\r') {
                uint8_t ret = '\n';
                HAL_UART_Transmit(&huart1, &ret, 1, HAL_MAX_DELAY);
        }

        return ch;
}
/* USER CODE END 4 */
```

4. Save and close the file.

5. Open app_fileX.c file.
6. In the "USER CODE BEGIN Includes" section, add the following:

```
/* USER CODE BEGIN Includes */
#include "main.h"
#include <stdio.h>
/* USER CODE END Includes */
```

7. In the "USER CODE BEGIN PV" section, define a FX_FILE type:

```
/* USER CODE BEGIN PV */
FX_FILE                    fx_file;
/* USER CODE END PV */
```

8. In the "USER CODE BEGIN 0" section add the following:

```
/* USER CODE BEGIN 0 */
printf("FX Initialization...\n");
/* USER CODE END 0 */
```

9. Now we move down to the main FileX thread. In the "USER CODE BEGIN fx_app_thread_entry 0" section, add the following:

```
/* USER CODE BEGIN fx_app_thread_entry 0 */
ULONG bytes_read;
ULONG available_space_post;
CHAR read_buffer[32];
CHAR data[] = "Hello from FileX on STM32\n";

printf("Staring FileX App Thread.\n");
/* USER CODE END fx_app_thread_entry 0 */
```

The code defines some variables that will be used to: read the file, define a string to send to the file, and create a variable to get the available space on the disk.

10. Finally, we will add the actual application that creates a file in the NOR flash disk, opens the file, writes to the file, and closes the file. Then the application will re-open the file, read the file, print the message to the console, and close the file. The final step is to get some information about the disk and close the NOR flash disk. In the "USER CODE BEGIN fx_app_thread_entry 1", add the following:

```
/* USER CODE BEGIN fx_app_thread_entry 1*/
  nor_ospi_status = fx_file_create(&nor_ospi_flash_disk,
"FXTEST.TXT");
    if (nor_ospi_status != FX_SUCCESS)
    {
        if (nor_ospi_status != FX_ALREADY_CREATED)
        {
```

```
                    /* Create error, call error handler.  */
                    Error_Handler();
                }
        }

    nor_ospi_status =  fx_file_open(&nor_ospi_flash_disk,
&fx_file, "FXTEST.TXT", FX_OPEN_FOR_WRITE);
    if (nor_ospi_status != FX_SUCCESS)
    {
      Error_Handler();
    }
    ////move the file pointer to the start of the file////////
    nor_ospi_status =  fx_file_seek(&fx_file, 0);
    if (nor_ospi_status != FX_SUCCESS)
    {
      Error_Handler();
    }
    /////////Write the message to the file and close the file/////
    nor_ospi_status =  fx_file_write(&fx_file, data,
sizeof(data));
    if (nor_ospi_status != FX_SUCCESS)
    {
      Error_Handler();
    }

    nor_ospi_status =  fx_file_close(&fx_file);
    if (nor_ospi_status != FX_SUCCESS)
    {

      Error_Handler();
    }

    /////Open the File, print the message to the console, and
close the file////////
    nor_ospi_status =  fx_file_open(&nor_ospi_flash_disk,
&fx_file, "FXTEST.TXT", FX_OPEN_FOR_READ);
    if (nor_ospi_status != FX_SUCCESS)
    {
      Error_Handler();
    }

    nor_ospi_status =  fx_file_seek(&fx_file, 0);
    if (nor_ospi_status != FX_SUCCESS)
    {
      Error_Handler();
    }

    nor_ospi_status =  fx_file_read(&fx_file, read_buffer,
sizeof(data), &bytes_read);
```

168

```
  if ((nor_ospi_status != FX_SUCCESS) || (bytes_read !=
sizeof(data)))
  {
    Error_Handler();
  }

  printf("Here is the file contents:  %s", read_buffer);


  nor_ospi_status =  fx_file_close(&fx_file);
  if (nor_ospi_status != FX_SUCCESS)
  {
    Error_Handler();
  }


  ////Gets some stats on the disk and close the disk///////
  nor_ospi_status =
fx_media_space_available(&nor_ospi_flash_disk,
&available_space_post);
  if (nor_ospi_status != FX_SUCCESS)
  {
    Error_Handler();
  }

  printf("Space available: %lu bytes\n", available_space_post);

  nor_ospi_status =  fx_media_close(&nor_ospi_flash_disk);
  if (nor_ospi_status != FX_SUCCESS)
  {
    Error_Handler();
  }

  printf("Leaving FileX App Thread.\n");
/* USER CODE END fx_app_thread_entry 1*/
```

11. Save and close the file.


## 11.8 Debug the Applications on the Board

We are ready to test the application on the board.

1. Build the application and make sure there are no errors.
2. In the app_filex.c file set a breakpoint at the line where the file is created:

```
181    ////////Create and open the file////////////
182    sram_status = fx_file_create(&sram_disk, "FXTEST.TXT");
183    if (sram_status != FX_SUCCESS)
```

3. Make sure the STM32U5 is connected to the development machine via the USB cable and start the debug session.
4. The debugger will stop at HAL_Init().
5. Open an Annabooks COM Terminal or equivalent serial terminal application.
6. Set the COM port to the STM32U5 USB COM port and initiate a connection.
7. Hit the debug continue button. The code will run until the breakpoint is hit.
8. Use the debug step-over function throughout the whole application. There is no access to looking at the flash, so the printf class helps track what is going on.

FX Initialization...
Staring FileX App Thread.
Here is the file contents: Hello from FileX on STM32
Space available: 67067904 bytes
Leaving FileX App Thread.

9. Stop the debugger.

From the output, the total available space in the disk is displayed, which we didn't get when using the RAM disk. Other disk functions can be performed, such as creating a directory, changing file names, deleting files, getting date and time information, setting attributes, etc. A list of these functions is in the fs_api.h file.

## 11.9 Summary: Storage – Details in the Setup

With these two FileX projects, the second guiding question has been addressed. The STM32Cube tools make it easy to get ThreadX running on different STM32 hardware platforms. The Nucleo-H723ZG RAM drive project made it easy to set up and view the results of the API calls in the Memory Monitor. Of course, data in RAM is lost when the system is powered down. Flash devices provide a persistent storage solution. In contrast to the RAM storage, the challenge is setting up the project in STM32CubeMX with all the little details for stack and pool sizes, as well as, configuring DMA channels correctly for flash devices. Once the project is set up, the application code is straight forward. These two projects used a single file as an example, but you could have multiple files in different directories to store data.

# 12 Project 8: Nucleo-H723 NetX Duo

Unlike the Discover boards, the Nucleo-H723 has a physical Ethernet port rather than wireless connectivity. The STM32H723 is based on the Cortex-M7 core, which has some extra features like a memory protection unit, MPU. This chapter looks at how to set up ThreadX for a physical Ethernet connection and enable the MPU.

## 12.1 *Create a Project with STM32CubeMX*

Unlike the STM32L4S5 Discovery Kit which had a wireless module connected to the SPI port, the physical Ethernet setup takes a little more work.

1. Open STM32CubeMX.
2. Under the New Project, click on "Access to Board Selector".
3. A new window appears. From the Commercial Part Number drop-down, select Nucleo-H723ZG. If you are using a different STM32 board, then select that product number.
4. On the right-hand side, there is a list of boards associated with the part number. There is only one on the list. Click on this item.

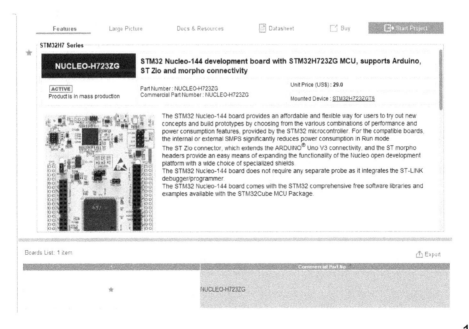

5. Click on "Start Project" in the top right corner.
6. You will be asked to initialize the default settings. Click Yes. The project gets initialized and a picture of the MCU appears with all the pins and associated I/O configuration for the development board already laid out. Since this is a development board, we will keep the defaults.
7. Now, we need to add Azure RTOS ThreadX and NetX Duo components to the project. Click on Software Packs->Select Components.
8. The package selector appears. There are some packages active and some that are inactive. The active packages are for the STM32 MCU on the board. The inactive packages are for other STM32 MCUs. Locate STMicroelectronics.X-CUBE-AZRTOS-H7 and click the Install button next to the package. This will install the Azure RTOS package support for the STM32 MCU that is on the board.
9. Once installed, expand the branches under RTOS ThreadX->ThreadX.
10. Tick the box next to Core.
11. Expand NetX Duo.
12. Tick the box next to NX Core.
13. Tick the box next to Addons DHCP Client.
14. Expand Network Interfaces.
15. Tick the box next to Ethernet interface.
16. In the Ethernet Phy interface drop-down, select LAN8472_Phys_Interface.
17. Expand Board Parts STM32_BSP_Components.
18. Tick the Box next to ETHPhy / LAB8472.

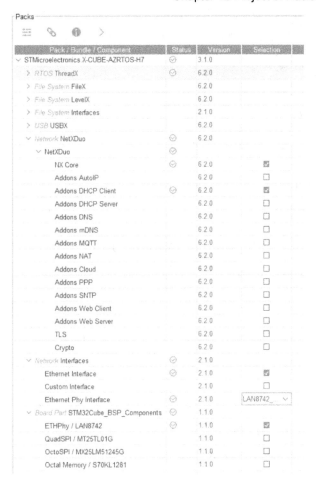

19. Click the OK button in the bottom right corner.
20. STMicroelectronics.X-CUBE-AZRTOS-H7 is now listed under Middleware and Software Packs. Click on STMicroelectronics.X-CUBE-AZRTOS-H7.
21. The STMicroelectronics.X-CUBE-AZRTOS-H7 Mode and Configuration will appear. Tick all the boxes for RTOS ThreadX, NetX Duo, and Network Interfaces.

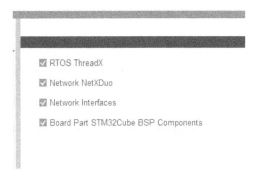

22. The configuration options are displayed below the Mode. The default NetX Duo memory pool size is too small, and the application will fail when a network connection is attempted. In the AzureRTOS Application tab, change the NetXDuo memory pool size to 30 * 1024.

23. Tick the box next to "Initialize DHCP protocol".

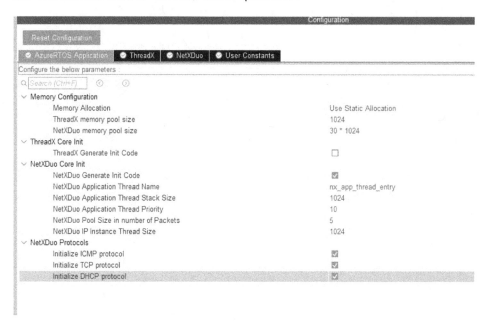

24. Under the NetXDuo tab, tick the boxes next to the following settings:

- NX_ENABLE_INTERFACE_CAPABILITY
- NX_DISABLE_IPV6

25. Under the categories on the left, expand Connectivity.
26. Click on ETH.
27. Make sure the Mode is set to RMII.
28. Change the Ethernet MAC Address to something unique.

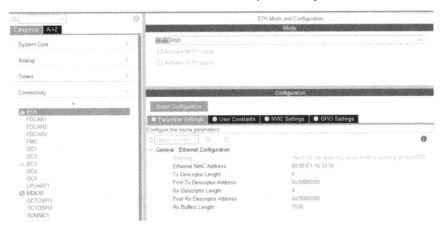

Like the physical flash chip in the last chapter, the physical ethernet needs a place in memory for transmit and receive data. The addresses provided in the settings will not have any effect on the project. Instead, the linker script will include additions to set the memory locations.

**Note**: If you were to start a project from the STM32H723 MCU, you would have to use the pin configuration to change two pins. Pins PG11 and PG13 would have to be added to replace the default implementation. The board schematic shows PG11 and PG13, which confirms this. When the pins are added, the default pins are then set to a reset state.

29. We need to configure the clock. Expand the System Core on the left side.
30. The SYS-tick is used by the HAL and Azure RTOS. To separate the two, we will give the HAL a different clock source. In the categories, select System Core->SYS.
31. In the Timebase Source drop-down, select TIM6.

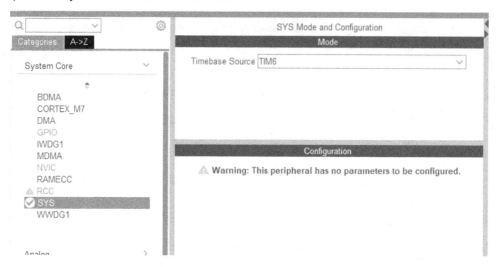

32. Now we need to configure the interrupts under System Core. Click on NVIC.
33. Tick the box next to the "Ethernet global interrupt".
34. Change the Preemption Priority for "Ethernet global interrupt" to 6
35. Change the Preemption Priority for TIM6 global interrupt to 0.

36. Now, let's enable the MPU settings. The MPU goal is to block code from accessing memory areas that it shouldn't access. A detailed discussion of the MPU is outside the scope of this book. STMicroelectronics has several videos and white papers discussing the MPU and setup details. Under System Core, click on CORTEX_M7.
37. Under Cortex interface Settings, Enable both CPU ICache and CPU DCache.
38. Under Cortex Memory Protection Unit Control Settings, set MPU Control Mode to "Background Region Privileged accesses only + MPU Disable during hard fault, NMI and FAULTMASK handlers".
39. The 16 MPU region settings become available. We want restrictions on the whole memory space and some sub-regions, but we want to open a region for the Ethernet TX / RX Descriptor addresses. Enter the following for Region 0:

- MPU Region: Enable
- MPU Region Base Address: 0x0
- MPU Region Size: 4GB
- MPU SubRegion Disable 0x87
- MPU TEX field level: level 0
- MPU Access Permission: All ACESS NOT PERMITTED
- MPU Instruction Access: DISABLED
- MPU Shareability Permission: ENABLED
- MPU Cacheable Permission: DISABLED
- MPU Bufferable Permission: DISABLED

40. For Region 1, set the following:

- MPU Region: Enabled
- MPU Region Base Address: 0x24030000
- MPU Region Size: 128KB
- MPU SubRegion Disable: 0x0
- MPU TEX field level: level 0
- MPU Access Permission: ALL ACCESS PERMITTED
- MPU Instruction Access: ENABLED
- MPU Shareability Permission: DISABLED
- MPU Cacheable Permission: ENABLED
- MPU Bufferable Permission: DISABLED

41. For Region 2, set the following:

- MPU Region: Enabled
- MPU Region Base Address: 0x24030000
- MPU Region Size: 256B
- MPU SubRegion Disable: 0x0
- MPU TEX field level: level 0
- MPU Access Permission: ALL ACCESS PERMITTED
- MPU Instruction Access: ENABLED
- MPU Shareability Permission: DISABLED
- MPU Cacheable Permission: DISABLED
- MPU Bufferable Permission: ENABLED

42. We will make a small clock adjustment. Click on the Clock Configuration.
43. Change DIVN1 to 260. This will lower the SYSCLK clock from the max value. Several examples have demonstrated this change from the default. The only effect that was seen in testing was a missed reply to the first ping request. Slowing the clock down, provided a full response.

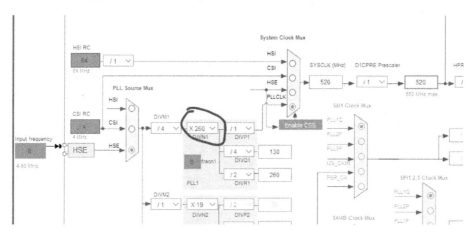

44. Click on the Project Manager tab.
45. Fill in the following:

    a. Project Name: Nucleo-H723-NetXDuo
    b. Project Location: Make the project location the same location for the STM32CubeIDE Workspace folder.
    c. Toolchain/IDE: STM32CubeIDE.
    d. Click on the "Generate Under Root" checkbox.

46. Click on "Generate Code" in the top right. The project will be created in the folder you selected.
47. Once the code has been generated, a dialog will ask you to open the files or open the project. Click on the "Open Project" button.
48. STM32CubeIDE will open and import the project into the workspace folder that you created in Chapter 3.
49. Close STM32CubeMX.

## 12.2 *Writing the Application.*

Like the NetX Duo for the Discovery board, the application simply enables the network stack so we can ping the board from another computer on the network.

1. Sending messages out the console port is helpful to track the operation of the NetX thread and get the IP address. In STM32CubeIDE, open main.h and add the prototype for PUTCHAR and GETCHAR to the "USER CODE BEGIN Private defines" section:

```
/* USER CODE BEGIN Private defines */
#define PUTCHAR_PROTOTYPE int __io_putchar(int ch)
#define GETCHAR_PROTOTYPE int __io_getchar(void)
/* USER CODE END Private defines */
```

2. Save and close the file.
3. Open main.c and add the putchar and getchar functions to the "USER CODE BEGIN 4" section:

```
/* USER CODE BEGIN 4 */
PUTCHAR_PROTOTYPE
{
/* Place your implementation of fputc here */
/* e.g. write a character to the USART1 and Loop until the end
of transmission */
HAL_UART_Transmit(&huart3, (uint8_t *)&ch, 1, 0xFFFF);

return ch;
}

GETCHAR_PROTOTYPE
{
        uint8_t ch;
```

```
    HAL_UART_Receive(&huart3, &ch, 1, HAL_MAX_DELAY);

    /* Echo character back to console */
    HAL_UART_Transmit(&huart3, &ch, 1, HAL_MAX_DELAY);

    /* And cope with Windows */
    if (ch == '\r') {
        uint8_t ret = '\n';
        HAL_UART_Transmit(&huart3, &ret, 1, HAL_MAX_DELAY);
    }

    return ch;
}
/* USER CODE END 4 */
```

4. In the "User CODE BEGIN Includes" section, add the following:

```
/* USER CODE BEGIN Includes */
#include <stdio.h>
/* USER CODE END Includes */
```

5. In the "User Code Begin 2" section, add the following printf call:

```
/* USER CODE BEGIN 2 */
printf("Staring the Threadx...\n");
/* USER CODE END 2 */
```

6. Save and close the file.
7. Open app_netxduo.c
8. In the "User CODE BEGIN Includes" section, add the following:

```
/* USER CODE BEGIN Includes */
#include "main.h"
#include <stdio.h>
/* USER CODE END Includes */
```

9. A couple of variables are needed to retrieve the TCP/IP address and NetMask. In the "USER CODE BEGIN PV" section add the following:

```
/* USER CODE BEGIN PV */
ULONG IpAddress;
ULONG NetMask;
/* USER CODE END PV */
```

10. Scroll down to the nx_app_thread_entry() function. In the "USER CODE BEGIN Nx_App_Thread_Entry 2" section, add the following to retrieve the TCP/IP address and print it to the console port:

```
nx_ip_address_get(&NetXDuoEthIpInstance, &IpAddress, &NetMask);
```

```
/* print the IP address and the net mask */
printf("IP address: %lu.%lu.%lu.%lu\r\n", (IpAddress >> 24),
(IpAddress >> 16 & 0xFF), (IpAddress >> 8 & 0xFF), (IpAddress &
0xFF));
```

11. Save and close the file.

Typically, you would want to add calls to error-handling functions when this function's return is not successful (ret != NX_SUCCESS). Adding code to handle errors is important for trouble shooting. If there are any issues with the network, the resulting TCP/IP address will be 0.0.0.0.


## 12.3 Modifying the Linker Script

The STM32L4S5 Discovery Kit used a Wireless chip that connected to the SPI port, and the buffering and access were handled in the SPI driver. For this physical Ethernet, we need to define input and output data space. When the code is generated for this project, there is some optional code put in to define the NetX Duo pool section, as well as, the TX/RX Descriptor addresses.

1. Open app_azure_rtos.c. In the Private variables section, you will see preprocessor code options for the particular compiler used, as well as, variables to define the memory address for the .NetXPoolSection.

```
#if defined ( __ICCARM__ )
#pragma data_alignment=4
#endif
__ALIGN_BEGIN static UCHAR
tx_byte_pool_buffer[TX_APP_MEM_POOL_SIZE] __ALIGN_END;
static TX_BYTE_POOL tx_app_byte_pool;

/* USER CODE BEGIN NX_Pool_Buffer */
#if defined ( __ICCARM__ ) /* IAR Compiler */
#pragma location = ".NetXPoolSection"
#elif defined ( __CC_ARM ) || defined(__ARMCC_VERSION) /* ARM
Compiler 5/6 */
__attribute__((section(".NetXPoolSection")))
#elif defined ( __GNUC__ ) /* GNU Compiler */
__attribute__((section(".NetXPoolSection")))
#endif
/* USER CODE END NX_Pool_Buffer */
#if defined ( __ICCARM__ )
#pragma data_alignment=4
#endif
__ALIGN_BEGIN static UCHAR
nx_byte_pool_buffer[NX_APP_MEM_POOL_SIZE] __ALIGN_END;
```

```
static TX_BYTE_POOL nx_app_byte_pool;
```

2. Close the file.
3. Open Main.c. In the Private variables section, you will see similar preprocessor code compiler options, as well as, variables defining the Ethernet TX / RX Descriptor (.TxDecripSection and .RxDecripSection).

```
#if defined ( __ICCARM__ ) /*!< IAR Compiler */
#pragma location=0x30000000
ETH_DMADescTypeDef  DMARxDscrTab[ETH_RX_DESC_CNT]; /* Ethernet
Rx DMA Descriptors */
#pragma location=0x30000200
ETH_DMADescTypeDef  DMATxDscrTab[ETH_TX_DESC_CNT]; /* Ethernet
Tx DMA Descriptors */

#elif defined ( __CC_ARM )  /* MDK ARM Compiler */

__attribute__((at(0x30000000))) ETH_DMADescTypeDef
DMARxDscrTab[ETH_RX_DESC_CNT]; /* Ethernet Rx DMA Descriptors */
__attribute__((at(0x30000200))) ETH_DMADescTypeDef
DMATxDscrTab[ETH_TX_DESC_CNT]; /* Ethernet Tx DMA Descriptors */

#elif defined ( __GNUC__ ) /* GNU Compiler */
ETH_DMADescTypeDef DMARxDscrTab[ETH_RX_DESC_CNT]
__attribute__((section(".RxDecripSection"))); /* Ethernet Rx DMA
Descriptors */
ETH_DMADescTypeDef DMATxDscrTab[ETH_TX_DESC_CNT]
__attribute__((section(".TxDecripSection")));   /* Ethernet Tx
DMA Descriptors */

#endif
```

4. Close the file.
5. These variables need to be defined in the linker script. Open STM32H723ZGTX_FLASH.ld.
6. Add the following towards the end of the file after the ._user_heap_stack and before the remaining discard section:

```
.tcp_sec (NOLOAD) :
{
  . = ABSOLUTE(0x24030000);
  *(.RxDecripSection)

  . = ABSOLUTE(0x24030060);
  *(.TxDecripSection)

} >RAM_D1 AT> FLASH
```

```
.nx_data 0x24030100 (NOLOAD):
{
    . = ABSOLUTE(0x24032100);
    *(.NetXPoolSection)
} >RAM_D1 AT> FLASH
```

7. Save the file.

The memory address of 0x24030000 is the same address defined in the MPU Region 1 and 2. The choice of using the 0x2403000 address in RAM_D1 was based on STM32H7 NetX Duo examples. The resulting code works. This is a bit of an issue. A developer might find this a little confusing as the STM32H7 datasheet suggests the address of 0x30000000 RAM_D2 for the Ethernet. The STM32 tools configure for RAM_D2 address 0x30000000 as we saw in the ETH settings. Finally, the FreeRTOS/ lwip examples show RAM_D2 being used. In our testing, we found that setting MPU regions 1 and 2 to address 0x30000000 and setting the linker file to the following worked:

```
.tcp_sec (NOLOAD) :
{
    . = ABSOLUTE(0x30000000);
    *(.RxDecripSection)

    . = ABSOLUTE(0x30000060);
    *(.TxDecripSection)

} >RAM_D2 AT> FLASH

.nx_data 0x30000200 (NOLOAD):
{
    . = ABSOLUTE(0x30000200);
    *(.NetXPoolSection)
} >RAM_D2 AT> FLASH
```

There is no impact on .elf file size. No explanation has been found as to why the examples are doing one thing versus the tools and datasheet recommending a different implementation. For now, we will use the original 0x2403000 address and linker settings.

## 12.4 Debug the Applications on the Board

We are ready to test the application on the board.

1. Build the application and make sure there are no errors.
2. Make sure the Nucleo-H723 is connected to the development machine via the USB cable and that the Ethernet is connected to a network that has a DHCP server. Start the debug session.
3. The debugger will stop at HAL_Init().
4. Open an Annabooks COM Terminal or equivalent serial terminal application.
5. Set the COM port to the Nucleo-H723 USB COM port and initiate a connection.

183

6. Hit the debug continue button. The code will run and it should display the TCP/IP address assigned by the DHCP server:

Staring the Threadx...
Starting NetX Duo
IP address: 192.168.1.33

7. Open a command prompt and send a ping to the address listed. You should get replies from the board.

C:\Users\SEAN_>ping 192.168.1.33

Pinging 192.168.1.33 with 32 bytes of data:
Reply from 192.168.1.33: bytes=32 time=1ms TTL=128
Reply from 192.168.1.33: bytes=32 time=1ms TTL=128
Reply from 192.168.1.33: bytes=32 time=1ms TTL=128
Reply from 192.168.1.33: bytes=32 time=1ms TTL=128

Ping statistics for 192.168.1.33:
    Packets: Sent = 4, Received = 4, Lost = 0 (0% loss),
Approximate round trip times in milli-seconds:
    Minimum = 1ms, Maximum = 1ms, Average = 1ms

8. Stop the debug session.


## 12.5 Summary: All the Small Things

There are a number of items to set up to get the Ethernet working. Any setting missed, incorrect address inputted, interrupt priorities set incorrectly, or pool size not big enough, will make the NETX Duo fail. Details of linker files and MPU settings are outside the scope of this book, but STMicroelectronics has a number of videos and white papers that provide different examples that are helpful. The next step would be to add support to connect to Azure, but we will leave this as an exercise for the reader.

# 13 Project 9: Dual-Core

All the boards we have covered so far are single-core MCUs. In this project, we will look at the STM32H747I-DISCO board that has two cores: Cortex-M7 and Cotrext-M4. Two cores allow two parallel workloads to be operating at the same time. Shared memory and Inter-Process Communication Control (IPCC) mechanisms allow for data to be exchanged between the two cores. One core can address real-time tasks and the other can address lower-priority tasks. With two large bus matrixes, the I/O can be configured for either core. Such a complex MCU could easily have several white papers to cover all the different capabilities and possibilities. All documentation has to start with the basics, so this chapter will focus on how to set up a project for both cores and configuring debugging so you can debug both cores interactively.

## 13.1 Create Dual-Core with STM32CubeMX

The project creates two subprojects, one for each core. There are 4 user LEDs available. Each core will simply flash one LED.

1. Open STM32CubeMX.
2. Under the New Project, click on "Access to Board Selector".
3. A new window appears. From the Commercial Part Number drop-down, select STM32H747I-DISCO.
4. On the right-hand side, there is a list of boards associated with the part number. There is only one on the list. Click on this item.

5. Click on "Start Project" in the top right corner.
6. You will be asked to initialize the default settings. Click Yes. The project gets initialized and a picture of the MCU appears with all the pins and associated I/O configuration for the development board already laid out. Since this is a development board, we will keep the defaults, but we will remove some items not being used.
7. Expand the categories. Notice that there are columns for each core or drop-downs to select a core.

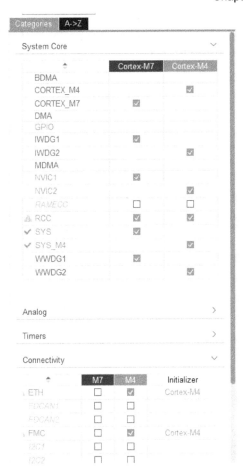

8. All the I/O and default configurations are not necessary for this project. Under Analog, uncheck all boxes for each core, VREFBUF cannot be changed.

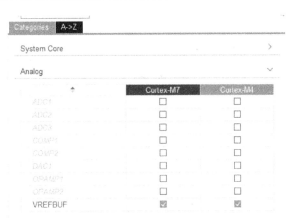

9. Under Timers, clear all checkboxes.
10. Under Connectivity, uncheck all boxes, except for USART1, which will be used by both cores. Leave the Cortex-M4 as the initializer for USART1.

| | | | |
|---|---|---|---|
| UART7 | ☐ | ☐ | |
| UART8 | ☐ | ☐ | |
| ⚠ USART1 | ☑ | ☑ | Cortex-M4 ⌄ |
| USART2 | ☐ | ☐ | |
| USART3 | ☐ | ☐ | |

11. Under Multimedia, uncheck all boxes.

Multimedia

| | Cortex-M7 | Cortex-M4 |
|---|---|---|
| DCMI | ☐ | ☐ |
| DMA2D | ☐ | ☐ |
| DSIHOST | ☐ | ☐ |
| HDMI_CEC | ☐ | ☐ |
| I2S1 | ☐ | ☐ |
| I2S2 | ☐ | ☐ |
| I2S3 | ☐ | ☐ |
| JPEG | ☐ | ☐ |
| LTDC | ☐ | ☐ |
| SAI1 | ☐ | ☐ |
| SAI2 | ☐ | ☐ |
| SAI3 | ☐ | ☐ |
| SAI4 | ☐ | ☐ |
| SPDIFRX1 | ☐ | ☐ |

12. Next, we need to set the Red LED to CM4 and the Green LED to CM7. Expand System Core and click on GPIO.
13. Under the GPIO tab, set PI14's Pin Context Assignment to ARM Cortex-M4.
14. Set PI12's Pin Context Assignment to ARM Cortex-M7.

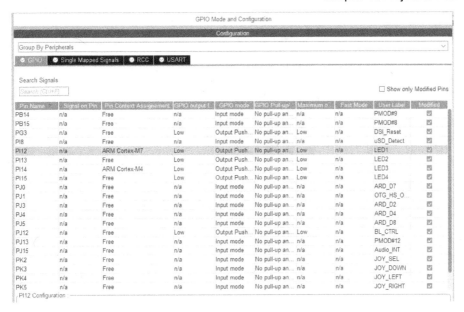

15. Click on the Project Manager tab.
16. Fill in the following:
    a. Project Name: H747DualCore.
    b. Project Location: Make the project location the same location as the STM32CubeIDE Workspace folder.
    c. Toolchain/IDE: STM32CubeIDE.
    d. Click on the "Generate Under Root" checkbox.
17. Click on "Generate Code" in the top right. The project will be created in the folder you selected.
18. Once the code has been generated, a dialog will ask you to open the files or open the project. Click on the "Open Project" button.
19. STM32CubeIDE will open and import the project into the workspace folder that you created in Chapter 3.
20. Close STM32CubeMX.

## 13.2 Modify the Main.c Files

Before we modify the code, let's take a look at the project structure.

1. Expand the new project.

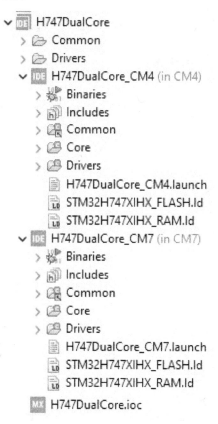

There are two sub-projects, one for each core that gets built independently. The default setup for the development board also sets up the HAL drivers. There is a common set of HAL drivers and then there are HAL drivers associated with each core that corresponds to the selections made in STM32CubeMX. Each core has its own linker description file and main.c file. We will refer to each core as CM4 and CM7, respectively. The startup sequence is a little different since there are two cores. The main.c file for each project has code to organize the synchronization. With the aid of a semaphore, CM4 will halt and wait for CM7 to send the activation call. Once the call has been made, CM7 halts. CM4 continues and releases the semaphore and continues the startup processing. With the semaphore released, CM7 will then start running again. Those are the differences between this project and our previous project. Now, we can move on with the code modifications for each subproject.

2. Open the main.h under the CM4 subproject.
3. Under the USER CODE BEGIN Private defines, add the follow:

```
#define PUTCHAR_PROTOTYPE int __io_putchar(int ch)
#define GETCHAR_PROTOTYPE int __io_getchar(void)
```

4. Save the file.

5. Open the main.c under the CM4 subproject.
6. Under USER CODE BEGIN Includes, add the stdio.h prototype:

```
/* USER CODE BEGIN Includes */
#include <stdio.h>
/* USER CODE END Includes */
```

7. Scroll down the User Code BEGIN 4 and add the following for UART debug output:

```
PUTCHAR_PROTOTYPE
{
    /* Place your implementation of fputc here */
    /* e.g. write a character to the USART1 and Loop until the end
of transmission */
    HAL_UART_Transmit(&huart1, (uint8_t *)&ch, 1, 0xFFFF);

    return ch;
}
```

```
GETCHAR_PROTOTYPE
{
        uint8_t ch;
        HAL_UART_Receive(&huart1, &ch, 1, HAL_MAX_DELAY);

        /* Echo character back to console */
        HAL_UART_Transmit(&huart1, &ch, 1, HAL_MAX_DELAY);

        /* And cope with Windows */
        if (ch == '\r') {
                uint8_t ret = '\n';
                HAL_UART_Transmit(&huart1, &ret, 1, HAL_MAX_DELAY);
        }

        return ch;
}
```

8. In main functions, add the following to USER CODE BEGIN 2

```
/* USER CODE BEGIN 2 */
printf("CM4 Running.\n");
/* USER CODE END 2 */
```

9. The RED LED has been set up for CM4. Add the following to the while-loop to flash the RED LED.

```
/* USER CODE BEGIN WHILE */
while (1)
```

```
{
```

```
        HAL_GPIO_TogglePin(LED3_GPIO_Port, LED3_Pin);
        HAL_Delay(1000);
```

```
    /* USER CODE END WHILE */
```

10. Save the file.
11. Open main.h for the CM7 subproject.
12. Under the USER CODE BEGIN Private defines, add the following:

```
#define PUTCHAR_PROTOTYPE int __io_putchar(int ch)
#define GETCHAR_PROTOTYPE int __io_getchar(void)
```

13. Save the file.
14. Open the main.c under the CM7 subproject.
15. Under USER CODE BEGIN Includes, add the stdio.h prototype:

```
/* USER CODE BEGIN Includes */
#include <stdio.h>
/* USER CODE END Includes */
```

16. Scroll down to the User Code BEGIN 4 section and add the following for UART debug output:

```
PUTCHAR_PROTOTYPE
{
  /* Place your implementation of fputc here */
  /* e.g. write a character to the USART1 and Loop until the end
of transmission */
  HAL_UART_Transmit(&huart1, (uint8_t *)&ch, 1, 0xFFFF);

  return ch;
}

GETCHAR_PROTOTYPE
{
        uint8_t ch;
        HAL_UART_Receive(&huart1, &ch, 1, HAL_MAX_DELAY);

        /* Echo character back to console */
        HAL_UART_Transmit(&huart1, &ch, 1, HAL_MAX_DELAY);

        /* And cope with Windows */
        if (ch == '\r') {
                uint8_t ret = '\n';
                HAL_UART_Transmit(&huart1, &ret, 1, HAL_MAX_DELAY);
```

```
    }

        return ch;
}
```

17. In main functions, add the following to USER CODE BEGIN 2

```
/* USER CODE BEGIN 2 */
MX_USART1_UART_Init();
HAL_Delay(100);
printf("CM7 Running.\n");
/* USER CODE END 2 */
```

18. The GREEN LED has been set up for CM7. Add the following to the while-loop to flash the LED.

```
/* USER CODE BEGIN WHILE */
while (1)
{

        HAL_GPIO_TogglePin(LED1_GPIO_Port, LED1_Pin);
        HAL_Delay(1000);

    /* USER CODE END WHILE */
```

19. Save the file.


## 13.3 Setting up the Debug Configuration for Both Projects

The steps so far have been straightforward for all the projects that we have done. The trick here is to start the debug session for two projects on two cores at the same time. The key to understanding why this is important can be seen in CM7's main.c. CM7 is considered CPU1 while CM4 is CPU2. On startup, CM7 is called first. As CM7 is initialized it will then turn over control to CM4 and wait. Once CM4 has initialized, both processors will run independently. The first part of this is to set up the debug configuration for each project.

1. Make sure that the board is <u>not</u> connected to the development machine.
2. Each sub-project is built separately. Right-click on the CM4 project, select "Build Project" from the context menu, and correct any errors.
3. Right-click on the CM7 project, select "Build Project" from the context menu, and correct any errors.
4. If there were any errors for either project, correct and rebuild.
5. Right-click on the CM7 project and select Debug AS->STM32 C/C++ Application.
6. The debug configuration dialog appears. Click on the Debugger tab.
7. Check the box next to Halt all cores. The rest of the settings can remain the same.

8. Click on the Startup tab.
9. The CM7 project is responsible for downloading and launching both cores. Click on the Add... button.
10. A dialog appears in the Project drop-down. Select H747DualCore_CM4.
11. Click OK when finished.

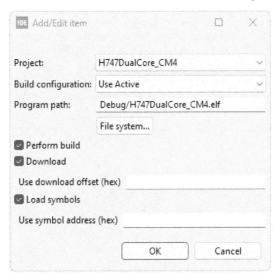

You should see both subprojects listed, and there is a green arrow next to the CM7 subproject, which indicates this is the project to start.

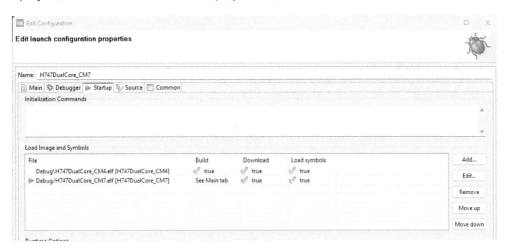

12. Click Apply.
13. Click Ok. The debugger will attempt to start but will fail as the board is not connected yet, but the configuration will be saved.
14. Right-click on the CM4 subproject and select Debug AS->STM32 C/C++ Application.
15. The debug configuration dialog appears.
16. Click on the Debugger tab.
17. Since the CM7 already has a port and is handling reset, set the following:
    a. Port number: 61238
    b. Reset behavior: None.

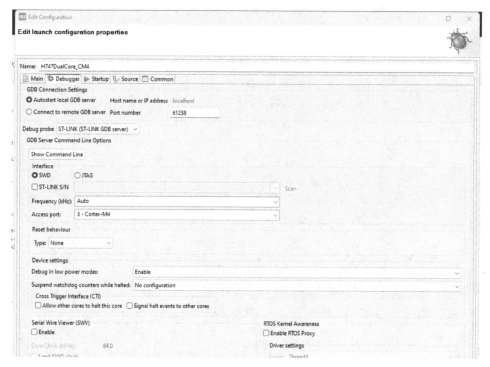

18. Click on the Startup tab.
19. Click on the line in the table for the CM4 subproject, and click the Edit... button.
20. Uncheck Download. The CM7 debug setup will already have downloaded this code.
21. Click OK when finished.

The debug configuration is now set to NOT download the CM4 project as the CM7 project will download both projects at once.

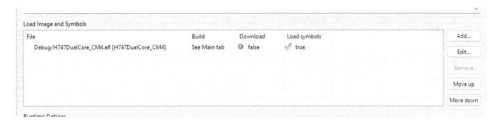

22. Click Ok.
23. Click Apply.
24. Click Ok. The debugger will attempt to start but will fail as the board is not connected yet, but the configuration will be saved.

## 13.4 Running the Debugger

The STM32Cube IDE supports debugging both cores in a single instance of the application. There is a little trick to running the debugger to debug both processors, which requires you to start the debug session for CM7, CPU1, first and then start a debug session for CM4, CPU2.

1. Make sure the STM32H747I-DISCO board is connected to the development machine.
2. Open the CM7's main.c and set a breakpoint at the call to MX_GPIO_Init() in the main() function.
3. Open the CM4's main.c and set a breakpoint at the call to MX_GPIO_Init() in the main() function.
4. Open an Annabooks COM Terminal or equivalent serial terminal application and connect the board's COM port.
5. Start the debugger for the CM7 project. Both elf files will be downloaded to the board. The board starts to run and stops at the timeout for CPU2.
6. Start the debugger for CM4. The debugger starts and stops at the semaphore clock enable.

**Note**: the console output has mixed messages about connected and waiting to connect to CM4. These can be ignored.

7. In the debug tree on the left, you will see that both cores have hit a breakpoint and have their respective active threads suspended. While holding down the control (CTRL) key, click on both main() functions to select them both.

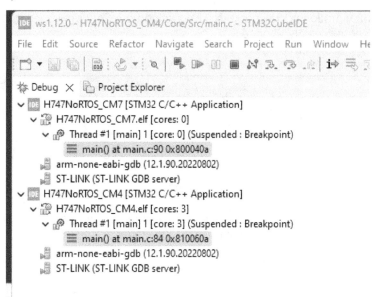

8. From the toolbar, click continue. Both cores will hit the breakpoints we added earlier.
9. In the main.c for the CM4, step through the code to toggle the RED LED.
10. In the debug tree, click on the main.c for the CM7. Step through the code to toggle the GREEN LED.
11. To stop debugging, while holding down the control (CTRL) key, select both main() threads, again in the debug tree, and click on stop debugging.

**Note**: Both cores have to be stopped to end both debug sessions and return to code view.

There are a few more tasks to be performed when working with two cores, but once you have the sequence figured out, the development process is not significantly different from a single-core MCU.

## 13.5 Add Azure RTOS to Both Cores

With the project set up and the ability to debug both cores covered, let's add ThreadX (Azure RTOS) to both cores and flash the LEDs in the threads.

### 13.5.1 Project Changes
We need to make some feature changes to the project to add Azure RTOS.

1. In STM32CubeIDE, open the H747DualCore.ioc file.
2. Expand the Timers Categories.
3. Set TIM6 to Cortex-M7 and TIM7 to Coretec-M4.

4. Expand the System Core category.
5. Click on SYS.
6. Set the Timebase Source to TIM6.

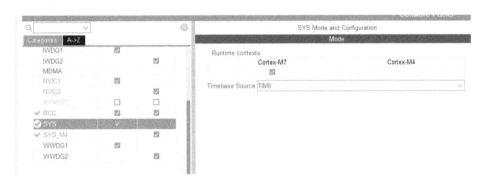

7. Click on SYS_M4.
8. Set the Timebase Source to TIM7.
9. Expand Connectivity.
10. Change the USART1 Initializer from Cortex-M4 to Cortex-M7.

| | | | |
|---|---|---|---|
| UART7 | ☐ | ☐ | |
| UART8 | ☐ | ☐ | |
| ⚠ USART1 | ☑ | ☑ | Cortex-M7 ⌄ |
| USART2 | ☐ | ☐ | |
| USART3 | ☐ | ☐ | |

11. Click on Software Packs->Select Components.
12. Since the MCU has two cores, the "Show components for context" appears at the top of the dialog. Select Cortex-M4.
13. Expand STMicroelectronics.X-CUBE-AZRTOS-H7.
14. Under RTOS ThreadX->ThreadX, check the box for Core.

| | | | |
|---|---|---|---|
| ⌄ STMicroelectronics.X-CUBE-AZRTOS-H7 | ⊘ | 3.1.0 ⌄ | |
| ⌄ RTOS ThreadX | ⊘ | 6.2.0 | |
| ⌄ ThreadX | ⊘ | | |
| Core | ⊘ | 6.2.0 | ☑ |
| PerformanceInfo | | 6.2.0 | ☐ |

15. Change "Show components for context" to Cortex-M7.
16. As was done for the Cortex-M4 core, expand STMicroelectronics.X-CUBE-AZRTOS-H7.
17. Under RTOS ThreadX->ThreadX, check the box for Core.
18. Click OK.
19. Expand the Middleware and Software Packs category.
20. There are two X-CUBE-AZRTOS-H7s. One is for each core. Click on the X-CUBE-AZRTOS-H7 for Cortex-M4.
21. Check the box next to "RTOS ThreadX".
22. In the Settings below under the AzureRTOS Application tab, check the box next to "ThreadX Generate Init Code".
23. For "Create ThreadX Application Thread" select True. When the project is saved a thread will be created to simply fill in the information.

24. In the categories, click on the X-CUBE-AZRTOS-H7 for Cortex-M7.
25. Check the box next to "RTOS ThreadX".
26. In the Settings below under the AzureRTOS Application tab, check the box next to "ThreadX Generate Init Code".
27. For "Create ThreadX Application Thread" select True. Again, a thread will automatically be set up and ready for coding.
28. Save the .ioc file and the new code will be generated.

### 13.5.2  Modify the Source Code
With the feature changes added, the next step is to add the code.

1. Open the app_threadx.c file for the H747DualCore_CM7 project.
2. Add the following includes:

```
/* USER CODE BEGIN Includes */
#include "main.h"
#include <stdio.h>
/* USER CODE END Includes */
```

3. Since all the work to set up a thread has been done, in tx_appThread_entry add the following:

```
while (1) {
        HAL_GPIO_TogglePin(LED1_GPIO_Port, LED1_Pin);
        HAL_Delay(1000);
        printf("CM7\n");
}
```

4. Save the file.
5. Build the CM7 project and correct any errors.
6. Open the app_threadx.c file for the H747DualCore_CM4 project.
7. Add the following include:

```
/* USER CODE BEGIN Includes */
#include "main.h"

/* USER CODE END Includes */
```

8. Add the following code to the tx_appThread_entry:

```
while (1) {

        HAL_GPIO_TogglePin(LED3_GPIO_Port, LED3_Pin);
        HAL_Delay(1000);

}
```

9. Save the file.
10. Build the CM4 project and correct any errors.

### 13.5.3  Debug the Code

The debug process is the same as before. Start the debug session for CM7; and once the breakpoint has been hit, start the debug session for CM4.

1. Make sure the STM32H747I-DISCO board is connected to the development machine.
2. Open the CM7's main.c and remove the breakpoint that was set earlier.
3. Open the CM7's app_threadx.c and set a breakpoint at the while-loop in tx_appThread_entry.
4. Open the CM4's main.c and remove the breakpoint that was set earlier.
5. Open the CM4's app_threadx.c and set a breakpoint at the while-loop in tx_appThread_entry.
6. Open ABCOMTerm or a similar serial terminal program and connect to the board's COM port.

7. Start the debugger for the CM7 project. Both elf files will be downloaded to the board. The board starts to run and stops at the time out for CPU2.
8. Start the debugger for CM4. The debugger starts and stops at the semaphore clock enable.

**Note**: the console output has mixed messages about connected and waiting to connect to CM4. These can be ignored.

9. In the debug tree on the left, you will see that both cores have hit a breakpoint and have their respective active threads suspended. While holding down the control (CTRL) key, click on both main() functions.
10. Both cores will run until they hit the breakpoints in the app_threadx.c files. You can now step through the code of each core separately.
11. To stop debugging, while holding down the CTRL key, use the mouse pointer to select both main() functions, again, and click on stop debugging.

## 13.6 Shared Memory Example

Having two cores allows for a division of work to help improve performance. Depending on the design of the device, there may be a need to exchange information between the two cores. STMicroelectronics has an application note that discusses the different cases for inter-processor communication between the cores. For this example, we will add a shared memory space between the two cores to facilitate the data transfer. Between the STM32H747 data sheet and IPCC application note (AN5617), most of the SRAM in the MCU can be shared between the two cores. The RAM domains for the 2 cores are as follows:

| Core | D1 Domain | | | D2 Domain | | | D3 Domain | |
|------|-----------|-----------|-------------|-----------|-----------|-----------|-----------|------------|
| | ITC M | DTC M | AXISRA M | SRAM 1 | SRAM 2 | SRAM 3 | SRAM 4 | BKSRA M |
| Cortex® -M7 | YES | | | Yes (cacheable) | | | | |
| Cortex® -M4 | Indirect (only via MDMA) | | | Yes | | | | |

The D1 Domain contains the Cortex-M7 core along with the AXI bus matrix. D2 Domain contains the Cortex-M4 core with the AHB bus matrix. The D3 Domain is used for system management and contains the SRAM4. More details about each domain and its uses are in the datasheets and application notes. For our purposes, the address locations for each SRAM are as follows:

AXISRAM is mapped at address 0x2400 0000
AHB SRAM1 is mapped at address 0x3000 0000 and 0x1000 0000
AHB SRAM2 is mapped at address 0x3002 0000 and 0x1002 0000
AHB SRAM3 is mapped at address 0x3004 0000 and 0x1004 0000
AHB SRAM4 is mapped at address 0x3800 0000

For this example, we will use SRAM4 to share memory between the two cores.

1. Open the_FLASH.ld file for the CM7 project.
2. Add the .sram4 to the SECTIONS section of the linker file:

```
SECTIONS
{

  .sram4 :
  {

    *(.sram4*);
  } > RAM_D3

  /* The startup code goes first into FLASH */
  .isr_vector :
  {
```

3. Save and close the file.
4. Open the_FLASH.ld file for the CM4 project.
5. Add RAM_D3 memory section:

```
/* Specify the memory areas */
MEMORY
{
    FLASH (rx)      : ORIGIN = 0x08100000, LENGTH = 1024K
    RAM (xrw)       : ORIGIN = 0x10000000, LENGTH = 288K
    RAM_D3 (xrw)    : ORIGIN = 0x38000000, LENGTH = 64K
}
```

6. Open the_FLASH.ld file for the CM7 project.
7. Add the .sram4 to the SECTIONS section of the linker file:

```
SECTIONS
{

  .sram4 :
  {

    *(.sram4*);
  } > RAM_D3

  /* The startup code goes first into FLASH */
  .isr_vector :
  {
```

8. Save and close the file.

9.  Open the app_thread.c for the CM4 project.
10. Add the following Private variable to use the shared memory space:

```
/* USER CODE BEGIN PV */
__attribute__((section(".sram4.counter"))) volatile unsigned int
counter;
/* USER CODE END PV */
```

11. Modify tx_app_thread_entry as follows:

```
void tx_app_thread_entry(ULONG thread_input)
{
  /* USER CODE BEGIN tx_app_thread_entry */
      while(1){

                HAL_GPIO_TogglePin(LED3_GPIO_Port, LED3_Pin);
                HAL_Delay(1000);
                counter++;
                //Reset counter when 100 is reached
                if(counter >= 100){
                    counter = 0;
                }

      }
  /* USER CODE END tx_app_thread_entry */
}
```

12. Save and close the file.
13. Open the app_thread.c for the CM4 project.
14. Add the following Private variable to use the shared memory space:

```
/* USER CODE BEGIN PV */
__attribute__((section(".sram4.counter"))) volatile unsigned int
counter;
/* USER CODE END PV */
```

15. Modify tx_app_thread_entry as follows:

```
void tx_app_thread_entry(ULONG thread_input)
{
  /* USER CODE BEGIN tx_app_thread_entry */

      while(1){

                HAL_GPIO_TogglePin(LED1_GPIO_Port, LED1_Pin);
                HAL_Delay(1000);
                printf("Counter Value: %u\n", counter);
      }
```

```
/* USER CODE END tx_app_thread_entry */
}
```

16. Save and close the file.

CM4 will increment the counter variable that is in the shared memory space. Once it hits 100, the counter value is set back to 0. CM7 will simply print the string with the counter value that is in memory at the time.

**Note**: Of course, the timing between each core might not make the output number sequential. Also, we have not enabled cache or the MPU to address cache coherency. This is just a simple demo to show shared memory access.

17. With the board connected to the development computer, open a terminal program to receive the debug messages.
18. Start the debug session for the CM7 project.
19. Once the CM7 has loaded and hits the breakpoint in main, start the debug session for the CM4 project.
20. Once the CM4 project has halted, select both main() functions in the Debug tree and click continue. Both of the projects will stop at their respective breakpoints. You can, now, step through the code for each processor.
21. Remove the breakpoints and select continue for both processors. The Debug output should show the counter incrementing and resetting to 0.

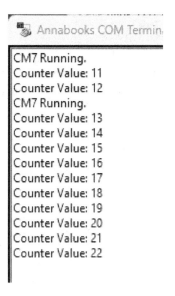

22. If you hit the reset button, the two cores will reset, but the value in SRAM remains.
23. Stop the debug session when you are finished.

## *13.7 Summary: Two Cores Better Than One*

A dual-core MCU offers a high-performance solution and a mechanism to divide up critical versus non-critical tasks between two cores. Developing with a dual-core system is different than a single-core MCU. The STM32Cube tools create a project for each core, since each core can run independently. Being able to debug both cores at the same time is a critical capability for multi-core development. The trick to debugging dual cores is in the debug configuration setup and the debug operation. After you have mastered the debug process, the rest of the development is straightforward. ThreadX (Azure RTOS) can be added to either or both cores, and we demonstrated inter-processor communication using shared memory. The next chapter will take a look at another IPCC solution.

# 14 Project 10: ThreadX and OpenAMP

In the last project, shared SRAM4 was used to share a counter variable between the two cores. Shared memory is one of the mechanisms the two cores can use to interact with each other. The problem with the last example is the fact that critical data could be missed by CM7 if CM4 is sped up. As we mentioned, if the timing were different, the output numbers wouldn't be sequential at all. Polling (loops) is not a good solution for communicating between cores. A better solution is to implement an interrupt call that lets CM7 know that data is ready when it arrives and to go and get it. This event-driven method removes the need for polling.

Rather than doing all the work to create your own event-driven message-passing IPC solution, there is already an open framework available in the repository. The OpenAMP (open asymmetric multi-processing) (https://www.openampproject.org/) is an open framework for created for the development of AMP systems. The OpenAMP software package is part of the STM32H7 repository and shows up as a middleware package in the .ioc file. The application note and OpenAMP website provide all the details of the library and its usage. The project in this chapter will be a modified version of an STMicroelectronics example project that will share a string between an Azure RTOS thread running on one core with another ThreadX thread running on the other core.

## 14.1 Create the OpenAMP Project with STM32CubeMX

We will create a brand-new project to test OpenAMP.

1. Open STM32CubeMX.
2. Under the New Project, click on "Access to Board Selector".
3. A new window appears. From the Commercial Part Number drop-down, select STM32H747I-DISCO.
4. On the right-hand side, there is a list of boards associated with the part number. There is only one on the list. Click on this item.

5. Click on "Start Project" in the top right corner.
6. You will be asked to initialize the default settings. Click Yes. The project gets initialized and a picture of the MCU appears with all the pins and associated I/O configuration for the development board already laid out. Since this is a development board, we will keep the defaults, but we will remove some items not being used.
7. Expand the categories. Notice that there are columns for each core or drops down to select a core.

8. Select Cortex_M7.
9. Enable the CPU ICache.
10. Enable the CPU DCache.
11. Now, we will enable the MPU to disable caching on the SRAM4. Under MPU Control Mode, select the 3rd option – "Background Region Privileged accesses only + MPU Disabled during hardware fault..."
12. Enable Cortex Memory Protection Unit Region 0 Settings.
13. Set the following:

   a. MPU Region base Address: 0x38000000
   b. MPU Region Size: 64KB
   c. MPU TEX field level: 0x0
   d. MPU Access Permission: ALL ACCESS PERMITTED
   e. MPU Instruction Access: Enabled
   f. MPU Shareability Permission: Enabled
   g. MPU Cacheable Permission: Disabled
   h. MPU Bufferable Permission: Disabled

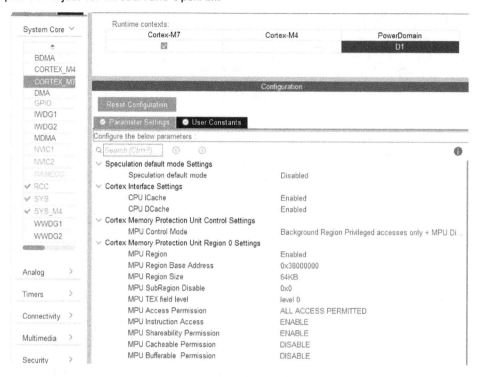

14. Click on GPIO.
15. Click on PI12 and set the "Pin Context Assignment" to ARM Cortex-M7.

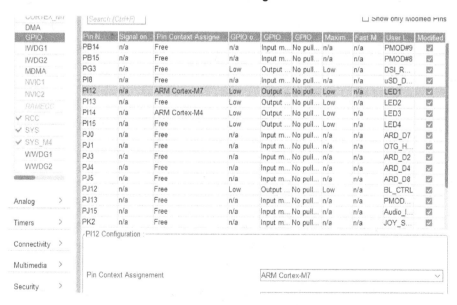

16. Click on PI14 and set the "Pin Context Assignment" to ARM Cortex-M4.
17. Click on NVIC1.

18. Check the box next to HSEM1 global interrupt. This will enable OpenAMP for Cortex-M7.

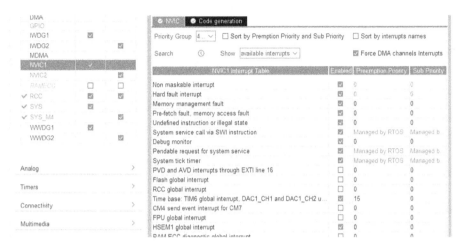

19. Click on NVIC2.
20. Check the box next to HSEM1 global interrupt. This will enable OpenAMP for Cortex-M4
21. All the I/O and default configurations are not required for this project. Under Analog, uncheck all boxes for each core, VREFBUF cannot be changed.

22. Under Timers, clear all checkboxes.
23. Check the box for TIM6 for Cortex-M7
24. Check the box for TIM7 for Cortex-M4

25. Under System Core, click on SYS.
26. Set "Timebase Source" to TIM6.
27. Click on SYS_M4.
28. Set "Timebase Source" to TIM7.

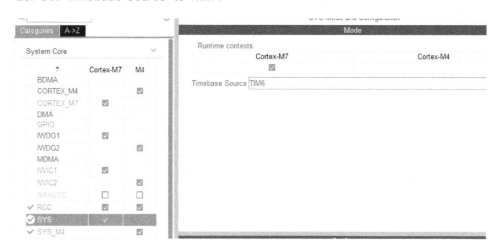

29. Under Connectivity, uncheck all boxes, except for USART1, which will be used by Cortex-M4. Leave the Cortex-M4 as the initializer for USART1.

| | Cortex-M7 | Cortex-M4 |
|---|---|---|
| UART7 | ☐ | ☐ |
| UART8 | ☐ | ☐ |
| ⚠ USART1 | ☐ | ☑ |
| USART2 | ☐ | ☐ |
| USART3 | ☐ | ☐ |

30. Under Multimedia, uncheck all boxes.

Multimedia

| | Cortex-M7 | Cortex-M4 |
|---|---|---|
| DCMI | ☐ | ☐ |
| DMA2D | ☐ | ☐ |
| DSIHOST | ☐ | ☐ |
| HDMI_CEC | ☐ | ☐ |
| I2S1 | ☐ | ☐ |
| I2S2 | ☐ | ☐ |
| I2S3 | ☐ | ☐ |
| JPEG | ☐ | ☐ |
| LTDC | ☐ | ☐ |
| SAI1 | ☐ | ☐ |
| SAI2 | ☐ | ☐ |
| SAI3 | ☐ | ☐ |
| SAI4 | ☐ | ☐ |
| SPDIFRX1 | ☐ | ☐ |

31. Under Middleware and Software Packs, click on OPENAMP_M4.
32. Check the box to Enabled. Keep all the default settings. The M7 will be the master and the M4 will be the slave.
33. Click on OPENAMP_M4.
34. Check the box to Enabled.

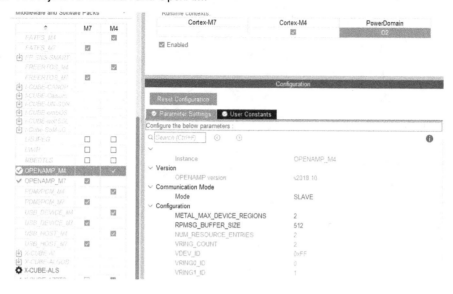

35. Click on Software Packs->Select Components.
36. Since the MCU has two cores, the "Show components for context" appears at the top of the dialog. Select Cortex-M4.
37. Expand STMicroelectronics.X-CUBE-AZRTOS-H7.
38. Under RTOS ThreadX->ThreadX, check the box for Core.

39. Change "Show components for context" to Cortex-M7.
40. Expand STMicroelectronics.X-CUBE-AZRTOS-H7.
41. Under RTOS ThreadX->ThreadX, check the box for Core.
42. Click OK.
43. Expand the Middleware and Software Packs category.
44. There are two X-CUBE-AZRTOS-H7s. One for each core. Click on the X-CUBE-AZRTOS-H7s for Cortex-M4.
45. Check the box next to "RTOS ThreadX".
46. In the Settings below, under the AzureRTOS Application tab, check the box next to "ThreadX Generate Init Code".
47. For "Create ThreadX Application Thread" select True. When the project is saved a thread will be created to simply fill in the information.

216

48. In the categories, click on the X-CUBE-AZRTOS-H7s for Cortex-M4.
49. Check the box next to "RTOS ThreadX".
50. In the Settings below, under the AzureRTOS Application tab, check the box next to "ThreadX Generate Init Code".
51. For "Create ThreadX Application Thread" select True. Again, a thread will automatically be set up and ready for coding.
52. Click on the Project Manager tab.
53. Fill in the following:
    e.  Project Name: H747AzureRTOSopenAMP.
    f.  Project Location: Make the project location the same location as the STM32CubeIDE Workspace folder.
    g.  Toolchain/IDE: STM32CubeIDE.
    h.  Click on the "Generate Under Root" checkbox.
54. Click on "Generate Code" in the top right. The project will be created in the folder you selected.
55. Once the code has been generated, a dialog will ask you to open the files or open the project. Click on the "Open Project" button.
56. STM32CubeIDE will open and import the project.
57. Close STM32CubeMX.

## 14.2 Modify the Source Code

The STMicroelectronics OpenAMP example is a bare metal no-RTOS solution. In fact, you will find similar solutions from other bloggers that cover bare-metal or FreeRTOS. For this project, the example will be integrated into ThreadX.

### 14.2.1 Apply a Patch

The example calls out a patch that must be implemented or will there will be errors in the build.

1. In STM32CubeIDE, open the errno.h file under Middlewares\Third_Party\OpenAMP\libmetal\lib\include\metal\compiler\mdk-arm.
2. Comment out the with in the ifndef __METAL_ERRNO__H__:

```
#ifndef __METAL_ERRNO__H__
//#error "Include metal/errno.h instead of metal/mdk-arm/errno.h"
#endif
```

3. Save and close the file.

**Note**: If you have to go back to the .ioc file and make project changes, the update will uncomment the line. You will have to comment out the line again for any project changes.

4. For both M4 and M7 sub-projects, open the respective syscalls.c files, add the following to undef errno after the includes:

```
#include <sys/time.h>
#include <sys/times.h>

#undef errno
extern int errno;
#include <sys/time.h>
```

5. Save and close the files.
6. For both M4 and M7 sub-projects, open the respective systmem.c files, add the following to undef errno after the includes

```
#include <errno.h>
#include <stdint.h>

#include <stddef.h>
#undef errno
extern int errno;
```

7. Save and close the files.

### 14.2.2 CM4 Subproject

The Cortex-M7 will send a message via the OpenAMP share to the Cortex-M4. The Cortex-M4 will receive the message and print the message to the ST-LINK COM port.

1. In STM32CubeIDE, open main.h.
2. Add the prototype for PUTCHAR:

```
/* USER CODE BEGIN Private defines */
#define PUTCHAR_PROTOTYPE int __io_putchar(int ch)
/* USER CODE END Private defines */
```

3. Save and close the file.
4. Open main.c.
5. Add the stdio.h include:

```
/* USER CODE BEGIN Includes */
#include <stdio.h>
/* USER CODE END Includes */
```

6. In "USER CODE BEGIN 4", add the following:

```
PUTCHAR_PROTOTYPE
{
    HAL_UART_Transmit(&huart1, (uint8_t*)&ch, 1, 0xFFFF);
    return ch;
}
```

7. In main(), add a message to indicate that the CM4 is running:

```
/* USER CODE BEGIN 2 */
printf("CM4\n");
/* USER CODE END 2 */
```

8. Save and close the file.
9. Open app_threadx.c.
10. Add the following includes:

```
/* USER CODE BEGIN Includes */
#include <stdio.h>
#include "main.h"
#include "openamp.h"
/* USER CODE END Includes */
```

11. Next, we will create a name for the RPMSG:

```
/* USER CODE BEGIN PD */
#define RPMSG_SERVICE_NAME              "openamp_demo"
```

219

```
/* USER CODE END PD */
```

12. Next, we will add the private variables:

```
/* USER CODE BEGIN PV */
static volatile int message_received;
volatile char *received_data_str;
static struct rpmsg_endpoint rp_endpoint;
/* USER CODE END PV */
```

13. Add the function prototype for the receive call-back and receive message:

```
/* USER CODE BEGIN PFP */
static int rpmsg_recv_callback(struct rpmsg_endpoint *ept, void
*data, size_t len, uint32_t src, void *priv);
unsigned int receive_message(void);
/* USER CODE END PFP */
```

14. Next, let's add the code that goes inside the thread:

```
void tx_app_thread_entry(ULONG thread_input)
{
  /* USER CODE BEGIN tx_app_thread_entry */
      int32_t status = 0;

      /* Initialize the mailbox use notify the other core on new
message */
      MAILBOX_Init();

      /* Initialize OpenAmp and libmetal libraries */
      if (MX_OPENAMP_Init(RPMSG_REMOTE, NULL) != HAL_OK) {
            Error_Handler();
      }

      /* Create an endpoint for rmpsg communication */
      status        =        OPENAMP_create_endpoint(&rp_endpoint,
RPMSG_SERVICE_NAME, RPMSG_ADDR_ANY, rpmsg_recv_callback, NULL);
      if (status < 0)
      {
        Error_Handler();
      }

      /* Receive a string from the master */
      receive_message();

      printf("String: %s\n\r", received_data_str);
```

```
            /* De-initialize OpenAMP */
            OPENAMP_DeInit();
    /* USER CODE END tx_app_thread_entry */
}
```

The code initializes OpenAMP and creates the end-point that sets up the receive callback function. The code then waits to receive a message from the master (CM7). Once the message has been received, the message is printed to the terminal and OpenAMP is de-initialized.

15. Finally, add the receive callback and receive message functions in the USER CODE BEGIN 1 section:

```
/* USER CODE BEGIN 1 */
static int rpmsg_recv_callback(struct rpmsg_endpoint *ept, void
*data,
                    size_t len, uint32_t src, void *priv)
{
    received_data_str = (char *) data;
    message_received=1;

    return 0;
}

unsigned int receive_message(void)
{
    while (message_received == 0)
    {
        OPENAMP_check_for_message();
    }
    message_received = 0;

    return 0;
}
/* USER CODE END 1 */
```

16. Save the file.
17. Build the CM4 project and correct any errors.


### 14.2.3  CM7 Subproject
CM7 is simply going to send a message.

1. Open in app_threadx.c in the CM7 subproject.
2. Add the following includes:

```
/* USER CODE BEGIN Includes */
#include "openamp.h"
#include "main.h"
/* USER CODE END Includes */
```

3. As with the CM4, project we have to match the channel name:

```
/* USER CODE BEGIN PD */
#define RPMSG_CHAN_NAME                      "openamp_demo"
/* USER CODE END PD */
```

4. Now, let's add the message string and the variables:

```
/* USER CODE BEGIN PV */
char str2cm4[] = "String from Azure RTOS Thread in CM7 to Azure
RTOS Thread in CM4 core using OpenAMP framework";
static volatile int message_received;
static volatile int service_created;
volatile unsigned int received_data_str;
static struct rpmsg_endpoint rp_endpoint;
/* USER CODE END PV */
```

5. Next, add the function prototypes:

```
/* USER CODE BEGIN PFP */
static int rpmsg_recv_callback(struct rpmsg_endpoint *ept, void
*data, size_t len, uint32_t src, void *priv);
unsigned int receive_message(void);
void service_destroy_cb(struct rpmsg_endpoint *ept);
void new_service_cb(struct rpmsg_device *rdev, const char *name,
uint32_t dest);
/* USER CODE END PFP */
```

6. Within the tx_app_thread_entry() add the following code:

```
/* USER CODE BEGIN tx_app_thread_entry */
        int32_t status = 0;

        /* Initialize the mailbox use notify the other core
on new message */
        MAILBOX_Init();

        /* Initialize the rpmsg endpoint to set default
addresses to RPMSG_ADDR_ANY */
        rpmsg_init_ept(&rp_endpoint,          RPMSG_CHAN_NAME,
RPMSG_ADDR_ANY, RPMSG_ADDR_ANY, NULL, NULL);
```

```
                /* Initialize OpenAmp and libmetal libraries */
                if  (MX_OPENAMP_Init(RPMSG_MASTER,  new_service_cb)!=
HAL_OK)
                {
                        Error_Handler();
                }


                /*
                * The rpmsg service is initiate by the remote
processor, on A7 new_service_cb
                * callback is received on service creation. Wait for
the callback
                */
                OPENAMP_Wait_EndPointready(&rp_endpoint);

                status    =    OPENAMP_send(&rp_endpoint,    str2cm4,
strlen(str2cm4) + 1);
                if (status < 0)
                {
                        Error_Handler();
                }

                /* Wait that service is destroyed on remote side */
                while(service_created)
                {
                        OPENAMP_check_for_message();
                }

                /* De-initialize OpenAMP */
                OPENAMP_DeInit();
  /* USER CODE END tx_app_thread_entry */
```

The code initializes the mailbox and configures RSPMSG to any RPMSG on the channel. The OpenAMP is initialized and then waits for the endpoint to be created. Once CM4 has created the endpoint, the message is sent to CM4. The thread waits for any messages from CM4 until the OpenAMP service has been de-initialized by CM4.In this case, there will be no messages from CM4, and CM4 just de-initializes the server and CM7 does the same.

7.   Finally, add the code behind the function prototypes:

```
/* USER CODE BEGIN 1 */
static int rpmsg_recv_callback(struct rpmsg_endpoint *ept, void
*data,
                    size_t len, uint32_t src, void *priv)
{
  received_data_str = *((unsigned int *) data);
```

```
    message_received=1;

    return 0;
}
unsigned int receive_message(void)
{
    while (message_received == 0)
    {
        OPENAMP_check_for_message();
    }
    message_received = 0;

    return 0;
}
void service_destroy_cb(struct rpmsg_endpoint *ept)
{
    /* this function is called when the remote endpoint has been
destroyed. The
     * service is no longer available
     */
    service_created = 0;
}

void new_service_cb(struct rpmsg_device *rdev, const char *name,
uint32_t dest)
{
    /* create an endpoint for rmpsg communication */
    OPENAMP_create_endpoint(&rp_endpoint,          name,          dest,
rpmsg_recv_callback,
                            service_destroy_cb);
    service_created = 1;
}
/* USER CODE END 1 */
```

8.  Save the file.
9.  Build the CM7 project and correct any errors.


## 14.3 Debug the Code

With the source code in place and two successful builds, we are ready to debug the project.

### 14.3.1 Debug Set Up

The debug setup will be the same as the last project.

1.  Make sure that the board is <u>not</u> connected to the development machine.
2.  Right-click on the CM7 project and select Debug AS->STM32 C/C++ Application.
3.  The debug configuration dialog appears. Click on the Debugger tab.

4. Check the box next to Halt all cores. The rest of the settings can remain the same.
5. Click on the Startup tab.
6. The CM7 project is responsible for downloading and launching both cores. Click on the Add... button.
7. A dialog appears in the Project drop-down, select H747AzureRTOSopenAMP_CM4.
8. Click OK when finished.

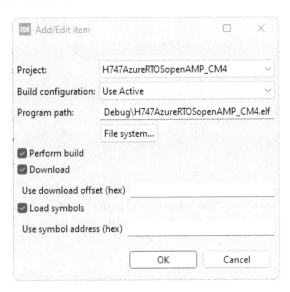

You should see both projects listed, and there is a green arrow next to CM7 project, which indicates this is the project to start.

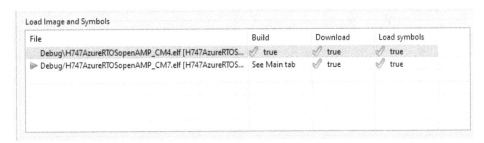

9. Click Apply.
10. Click Ok. The debugger will attempt to start but will fail as the board is not connected yet, but the configuration will be saved.
11. Right-click on the CM4 project and select Debug AS->STM32 C/C++ Application.
12. The debug configuration dialog appears.
13. Click on the Debugger tab.
14. Since the CM7 project already has a port and is handling reset, set the following:
    a. Port number: 61238.
    b. Reset behavior: None.

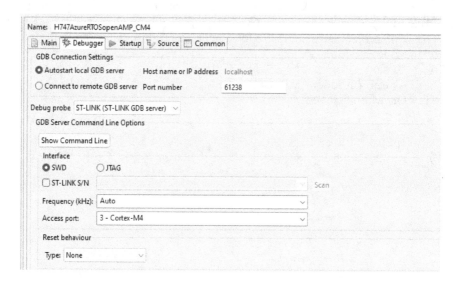

15. Click on the Startup tab.
16. Click on the line in the table for the CM4 project and click the Edit… button.
17. Uncheck Download. The CM7 debug setup will have already downloaded this code.
18. Click OK when finished.

The debug configuration is now set to NOT download the CM4 project as the CM7 project will download both projects at once.

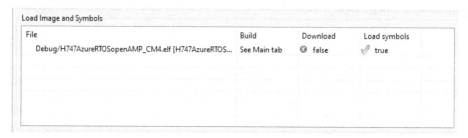

19. Click Ok.
20. Click Apply.
21. Click Ok. The debugger will attempt to start but will fail as the board is not connected yet, but the configuration will be saved.

### 14.3.2  Running the Debugger
Finally, we are ready to test the code

1. Make sure the STM32H747I-DISCO board is connected to the development machine.

2. Open a terminal program and connect the board's COM port.
3. Start the debugger for the CM7 project. Both elf files will be downloaded to the board. The board starts to run and stops at the time out for CPU2.
4. Start the debugger for CM4. The debug starts and stops at the semaphore clock enable.

**Note**: the console output has mixed messages about connected and waiting to connect to CM4. This can be ignored.

5. In the debug tree on the left, you will see that both cores have hit a breakpoint, and both threads are suspended. While holding down the control (CTRL) key, click on both main() functions to select them both.

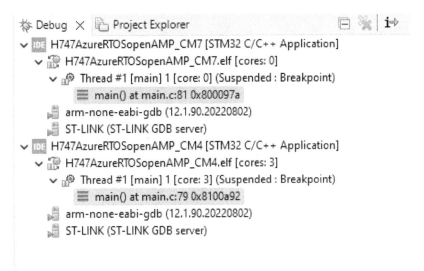

6. From the toolbar, click continue. The message should appear on the terminal.

7. To stop debugging, select both main() threads again in the debug tree and click on stop debugging.

**Note**: Both cores have to be stopped to stop debugging.

## 14.4 Send data from Slave to Master

The project simply sends a string from the master to the slave, and then both threads exit. Now, we will modify the code to have the slave send data to the master and toggle an LED. Both threads will have an infinite loop to keep the threads alive and the LED toggling.

1. In the CM4 project, open app_threadx.c
2. After the printf to send the message to the terminal, add the following and comment out the DeInit:

```
        int ledON = 0;
        int ledOFF = 1;


        while(1){

                status = OPENAMP_send(&rp_endpoint, &ledON,
sizeof(ledON));
                if (status < 0)
                {
                        Error_Handler();
                }
                HAL_Delay(1000);

                status = OPENAMP_send(&rp_endpoint, &ledOFF,
sizeof(ledOFF));
                if (status < 0)
                {
                        Error_Handler();
                }
                HAL_Delay(1000);
        }

    /* De-initialize OpenAMP */
//        OPENAMP_DeInit();
    /* USER CODE END tx_app_thread_entry */
```

Two integer messages are defined, one for turning on the LED and the other for turning off the LED. The message to turn on is sent first followed by a message to turn off after a delay. The sequence repeats forever.

3. Save the file.
4. Build the CM4 project and fix any errors.
5. In the CM7 project, open app_threadx.c.
6. After the sending of the message to CM4, add the following and comment out the service check and DeInit:

```
                //continue to receive data from slave to toggle LED
                while(1){

                        receive_message();

                        if(received_data_str == 0 ){

                                HAL_GPIO_TogglePin(LED1_GPIO_Port,
LED1_Pin);

                        }else
                        {
                                HAL_GPIO_TogglePin(LED1_GPIO_Port,
LED1_Pin);
                        }
                }

                /* Wait that service is destroyed on remote side */
//              while(service_created)
//              {
//                      OPENAMP_check_for_message();
//              }

                /* De-initialize OpenAMP */
//              OPENAMP_DeInit();
    /* USER CODE END tx_app_thread_entry */
```

7. Save the file.
8. Build the CM7 project and fix any errors.
9. Run the debugger. The message will be sent to the terminal as before, and the GREEN LED will be toggling on and off.

## 14.5 *Summary: Dual-Cores Working Together*

The STM32H747I-DISCO dual-core platform offers a lot of capability both in features and what the MCU offers, which will not be covered in this text. These last two chapters have introduced the idea of a dual-core MCU where the two cores can communicate with each other. The previous chapter had a simple solution for sharing data over shared memory space, but OpenAMP provides a more robust structured solution. This chapter's project was a simple example. A more complex solution would have had the M7 core handling the LCD and receiving data from the M4 core while the M4 core processes data in real time from all the I/O channels. The next chapter will look at displaying something on the LCD display.

# 15 Graphical User Interface Introduction

All the previous chapters have addressed the two guiding questions for the book. Along the way, the different features of ThreadX such as: ThreadX, TraceX, FileX, NetX Duo, and the ability to support a multicore system have been introduced. This leaves two ThreadX features not covered so far: USBX and GUIX. USBX will not be covered. A deep dive into GUIX is beyond the scope of the book, but what was interesting is why GUIX is not a selectable software package in the STM32Cube tools. Since we already had a system with LCD hardware, some time was dedicated to research the GUIX topic. The subtle issues with procedural steps and little details were enough to warrant including this chapter so other developers would not run into the same issues. This chapter serves as a light introduction to GUIX and TouchGFX.

## 15.1 Where is GUIX in STM32Cube and What is TouchGFX?

You probably have noticed that there are selectable ThreadX (Azure RTOS) software packages for ThreadX, TraceX, FileX, USBX, and NetX DUO that can be added to a project. There is no option for GUIX. There is a simple business reason why GUIX was excluded. STMicroelectronics has invested in the development of its own GUI framework library called TouchGFX. In addition, they have developed a design tool, as well as, many examples and training labs for TouchGFX. From a business perspective, the large investment into TouchGFX is something that STMicroelectronics doesn't want to throw away. Finally, it is incumbent on STMicroelectronics to provide support for ThreadX on their STM32 MCUs, GUIX is not something they want to support. Regardless of the exact reason, both GUI solutions will be introduced in this chapter to some level.

### 15.1.1 GUIX and TouchGFX Similarities
Both solutions have some similarities.

- Both GUI solutions have development tools to create the GUI application. GUIX has GUIX Studio. TouchGFX has TouchGFX Designer.
- The development tools allow you to run/test the GUI in Windows.
- Learning a new API can be challenging. Both GUI solutions have examples that are very important to study, so you can implement them in your applications.
- Code generation takes the graphics and puts them into code (.c or .cpp) files.

### 15.1.2 GUIX

Since ThreadX runs on different MCUs, GUIX Studio is a generic and very bare-bones development tool. Here are the advantages and disadvantages of GUIX:

- GUIX was designed to work with ThreadX. TouchGFX has some issues here.
- GUIX output files are C-language, which works directly with the C code in ThreadX. The C output files are easy to understand for making code changes.
- As mentioned, you can run GUIX applications in Windows, but it is not just a simulation. ThreadX has a port to Windows. You can build a GUIX application that is a Windows application. The real advantage is that once you have tested the GUI on Windows, you can move the GUI to an MCU running ThreadX. There are, however, some configuration steps that must be followed when moving the code to ThreadX on an MCU.
- GUIX requires extra work to integrate into an STM32CUBEIDE project. The GUIX Studio output files can be directly copied, but you have to designate memory for the framebuffer, add code to handle touchscreen input, add code to update the framebuffer, and configure MPU settings.
- During the installation of GUIX Studio 6.2.1.4, the GUIX repository couldn't be downloaded. A manual download was required.
- There are example projects centered around the GUIX API but no specific examples of how to integrate them into an MCU. One has to rely on the MCU vendor to provide support or do extensive searches online.

### 15.1.3 TouchGFX

TouchGFX was designed to work with the STM32CUBE and other MCU development tools. There are some advantages and disadvantages:

- TouchGFX Designer creates an STM32CubeIDE project for a development kit. No extra work to integrate the GUI into STM32CubeIDE is required. All the little details for the framebuffer, touch input, and memory configuration are already set up, but there are some problems with the TouchGFX STM32CubeIDE generated project.
  - The project structure is not the same as the project structure generated by STM32CubeMX. Include files are not easily accessible in the Project Explorer as they are from an STM32CubeMX-generated project. Also, the support for IAR and Kiel tools is in the file directory structure, which can and should be removed.
  - After testing the TouchGFX app on two platforms, there were some consistency issues. The same GUI ran perfectly on one development board but had issues with a different development board.
  - The STM32CubeMX .ioc file provided by the TouchGFX project is a version behind the current version. A migration to the latest version can result in some error messages and unknown changes to the project code when the file is saved, which can result in a broken build.
  - When creating a project based on a specific development board, the TouchGFX project name doesn't translate into the name of the project in STM32CubeIDE. Instead, the name of the development board is used,

which can be problematic when you want to create different TouchGFX application projects for the same board.

- TouchGFX Designer has demo and example projects that are easily downloaded and used to study the API. A project can also be created based on one of the many STM32 development boards that are available. The only issue is that you need to address any board revision changes. For example, the STM32H747I-DISCO board uses the latest MB1166-A09 daughter board, which has a different LCD controller chip. The project has to be adjusted to the latest LCD controller chip.
- A project can be created from STM32CuibeMX and the TouchFGX software package can be added independently.
- When selecting an STM32 development board from which to start a project, you will notice that most of the projects will set up FreeRTOS to run the TouchGFX application. As of this writing, only two boards can be created with ThreadX running TouchGFX.
- TouchGFX creates C++ files and has a more complex application structure that the uninitiated would have to spend some time to figure out. Any calls to C functions require the "extern C" pre-processor directives. Some extra coding is required for C language function calls to C++ language functions and vice-versa.
- TouchGFX offers some nice built-in animation for different widgets.
- There are differences between the GUIX TextBox widget and the TouchGFX TextArea widget. The GUIX TextBox widget's capability is more in line with standard GUI programming styles. TouchGFX TextArea widget requires extra thinking to address dynamic changes during the running of an application .

TouchGFX sounds like a good choice, since it integrates well with the rest of the STM32CUBE development tools; but there is a learning curve to understand the intricacies of the project structure, API, and limitations of the API. In the end, you will get the GUI up and running quickly on your system. GUIX is simpler but takes more work to integrate into the development tools.

## 15.2 Project 11 Create Project in TouchGFX Designer

In this first project, TouchGFX Designer will be used to create a simple GUI application for the STM32H747I-DISCO board. Chapter 3 had you download and install the TouchGFX Designer tool.

### 15.2.1 Create the IDE in TouchGFX Desinger
The application will have a button that toggles the text in a text area widget.

1. Open TouchGFX Designer.
2. The home screen appears. There are three buttons on the side: Create a project for reference board, create a new project based on an already existing example, and full demo project. Click on the Create New button.
3. You will be dropped down into the Create section that list all the currently support STM32 boards. Search for and select the STM32H747I-DISCO board.

4. Click the download arrow. The system will download some files to the computer and replace the download icon with a storage icon.
5. With the STM32H747I-DISCO selected, in the bottom right corner enter H747-TouchGFX for the application name.

6. The project is created, and you will get a blank canvas. From the menu, select code open files. File Explorer opens and you see the location of the TouchGFX project under the CM7 sub-project. The folder contains the H747-TouchGFX.touchgfx project file that can be re-opened in the future.

H747-TouchGFX › CM7 › TouchGFX

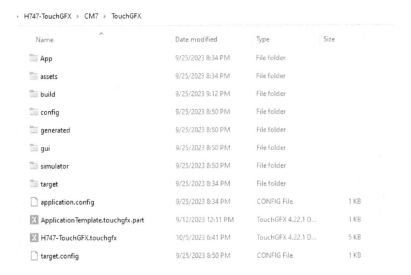

7. Go to the H747-TouchGFX folder. This is the whole project for the board that gets created. The STM32CubeMX .ioc file is part of the root. Also, the project files for IAR, Kiel, and STM32CubeIDE are in their respective folders.

H747-TouchGFX

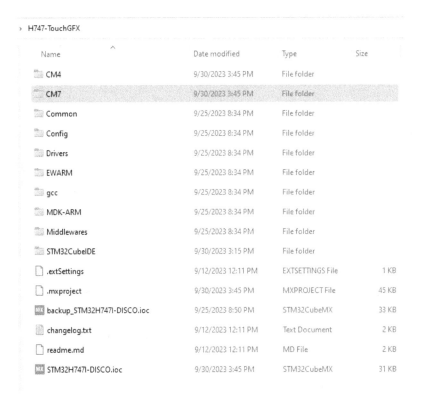

8. Close File Explorer and go back to TouchGFX Designer.
9. The toolbar above the canvas contains the different design elements that can be put into the design. Click on the shapes icon, and select box.

10. A box will be placed on the canvas. Expand the box so it fills the whole canvas.
11. On the right side change the color to something darker. This will be the background for the application.
12. From the toolbar, under buttons, select "Button with Label".
13. The button is placed on the canvas. Move the button to the top middle of the application.
14. In the properties on the right, let's fill in some information:
     a. Button Label: btnHello
     b. Text ID: btnHelloText
     c. Text Translation: Click ME
     d. Text Alignment: centered

**Properties**    Interactions

☐   btnHello

Location

X  258        Y  92

W  240        H  50

☐ Lock

✓ Visible

Preset

— Medium Rounded Normal

Image

Released Image

— medium_rounded_normal.png

Pressed Image

— medium_rounded_pressed.png

Text

ID

btnHelloText        ✕  🔍

Translation

Click Me

Typography

Default, 20px

Alignment

15. From the toolbar, under miscellaneous, select "Text Area".
16. The Text Area widget is placed in the application. Move the Text Area widget to the center of the application under the button.
17. In the properties, enter the name txtArea and uncheck Auto-size.

18. Change the font color to White 0xFCFCFC.
19. Under Text, delete the text and click on the + icon button to add a Wildcard1.
20. The Widlcard1 dialog appears, click the + icon and set the initial Value to "Click the button"
21. Check the "Use wildcard buffer" and set the Buffer size to 20.

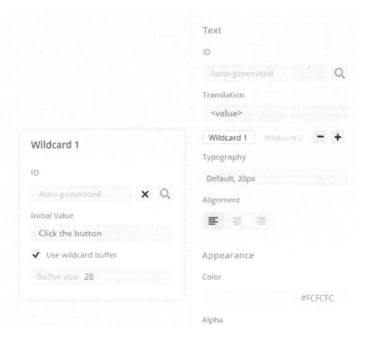

22. Click anywhere on the canvas to close the dialog.

238

23. Click on the textArea widget and use the handle to expand the width.

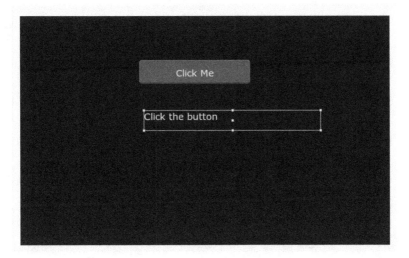

24. The textArea displays static text. Wildcards are used to dynamically allow the text in the textArea to be changed. This is one of the subtle differences that is unique to TouchGFX. Since we added the wildcard, we need to set the ASCII table range that the wildcard can use. On the left side, click on the Texts icon.
25. Click Typographies.
26. A dialog will appear for the Default Typography. Enter 0x20-0x7E, which lists all the lower case letters, upper case letters, and numbers from the ASCII table.

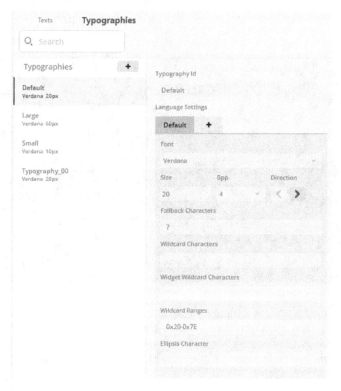

27. Click on the Canvas icon on the left to go back to the canvas.
28. Click on the Interactions in the top right.
29. Click the + icon.
30. A dialog will appear. Enter the following:
    a.  Trigger: Button is clicked
    b.  Choose clicked source: btnHello
    c.  Action: Call new virtual function
    d.  Function Name: toggle
    e.  Interaction Name: toggleText

This sets up the action and function behind the button, but we still have to write the code for the button event handler. We will do this in STM32CubeIDE.

31. From the menu, select File->Save.
32. From the menu, select Code->Generate Code.
33. If you want to run the application in the simulator, from the menu, select Code->Run Simulator. You can click on the button and nothing happens. Close the simulator when finished.
34. From the menu, select Code->Open Files.
35. Go to the ~\H747-TouchGFX\STM32CubeIDE folder and double-click on the .project file. This will open STM32CubeIDE and import the project.
36. Close TouchGFX Designer.

### 15.2.2 Edit the Code

STM32CubeIDE is used to complete the project. Now that the project has been imported you can see that the project name takes on the name of the development board and not the TouchGFX project name. This is not ideal and hopefully a bug that will be fixed in the future. There is a way to manually edit he project files to change the name, but this is a little bit cumbersome. For a custom board, STM32CubeMX and adding the TouchGFX software package is probably the best-known practice to create and work with STM32 projects.

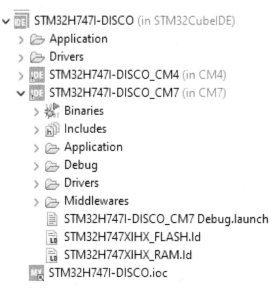

1. Under the CM7 sub-project, expand the branches of Application\User\Core.
2. Open the main.c file.
3. Since editing the STM32CubeMX .ioc file will make some changes, some code needs to be moved so it doesn't get lost during the .ioc migration. Move the app_touchgfx.h to the USER CODE BEGIN includes:

```
/* USER CODE BEGIN Includes */
#include "app_touchgfx.h"
/* USER CODE END Includes */
```

4. Scroll down to the main() function and move MX_TouchGFX_Init() and MX_TouchGFX_PreOSInit() to the USER CODE BEGIN 2 section.

```
/* USER CODE BEGIN 2 */
MX_TouchGFX_Init();
/* Call PreOsInit function */
MX_TouchGFX_PreOSInit();
/* USER CODE END 2 */
```

5. Save and close the file.

**Note**: The housekeeping that we just performed in main.c was unique to the STM32H747I-DISCO project. In comparison, a TouchGFX project for the STM32F769I-DISCO board didn't require these changes.

6. Let's add the code behind the button to toggle the textArea text. Under \Application\User\generated, open Screen1ViewBase.cpp. This file has some

information that we need in the other files, such as the name of the txtArea buffer. We can see the button handler call and the toggle() function when btnHello is clicked.

7. Under \Application\User\gui, open the Screen1View.cpp file and add the following code after the tearDownScreen() function:

```
void Screen1View::toggle(){

    if(toggleFlag == 0){
        Unicode::UnicodeChar buffer[20];
        const char str[] = "Hello World";
        Unicode::strncpy(buffer, str, 20);
        Unicode::snprintf(txtAreaBuffer, TXTAREA_SIZE, "%s",
buffer);
        txtArea.setWildcard1(txtAreaBuffer);
        txtArea.invalidate();
        toggleFlag = 1;
    }
    else
    {
        Unicode::UnicodeChar buffer[20];
        const char str[] = "TouchGFX Works!";
        Unicode::strncpy(buffer, str, 20);
        Unicode::snprintf(txtAreaBuffer, TXTAREA_SIZE, "%s",
buffer);
        txtArea.setWildcard1(txtAreaBuffer);
        txtArea.invalidate();
        toggleFlag = 0;
    }
}
```

The toggle() function is called by the event handler when btnHello is clicked. Based on a toggle value, the text will be changed for the txtArea and the screen will be refreshed. There is a bit of a trick to change the code. An array is created and a string is copied into the array, which gets rid of any end-of-line characters. The array is then copied into the txtAreaBuffer. The names txtAreaBuffer and TXTAREA_SIZE were taken from the Screen1ViewBase.cpp file. A call is then made to fill in the wildcard characters that match the text, which is the alphabet in this case. Finally, the invalidate() function is called to refresh the screen and the toggle flag is set.

8. Save the file.
9. We still need to add the declarations for the toggle function and the toggleFlag. Right-click on the #include <gui/screen1_screen/Screen1View.hpp> and select "Open Declaration".
10. The Screen1View.hpp file is opened. Add the virtual void toggle() under public and the toggleFlag integer under private.

243

```
#ifndef SCREEN1VIEW_HPP
#define SCREEN1VIEW_HPP

#include <gui_generated/screen1_screen/Screen1ViewBase.hpp>
#include <gui/screen1_screen/Screen1Presenter.hpp>

class Screen1View : public Screen1ViewBase
{
public:
    Screen1View();
    virtual ~Screen1View() {}
    virtual void setupScreen();
    virtual void tearDownScreen();
    virtual void toggle();

protected:

private:
    uint8_t toggleFlag = 0;

};

#endif // SCREEN1VIEW_HPP
```

11. Save and close the file.
12. There was a revision change to the MBII66 LCD daughter board. A new controller was added. We need to add the new controller driver files and switch the code over. Open File Explorer and open the folder location: Repository\STM32Cube_FW_H7_V1.11.1\Drivers\BSP\Components.
13. Copy the nt35510 folder and paste the folder to the C:\TouchGFXProjects\H747-TouchGFX\Drivers folder.
14. In STM32CubeIDE, under the CM7 sub-project, expand the Drivers\BSP\Components folder.
15. Right-click on Components and select Import.
16. With File System highlighted, click Next.
17. Click the Browse button and navigate to the C:\TouchGFXProjects\H747-TouchGFX\Drivers folder and select Open.
18. The import dialog shows the files in the folder. Select all the .C and .H files and click Finish.

19. Refresh the project. Right-click on the Components folder and select Add/remove include paths.
20. The dialog appears. Make sure both items are checked and select OK. This will allow the build to find the two .h files.
21. In the CM7 sub-project, open main.c.
22. Add the includes for the driver:

```
/* USER CODE BEGIN Includes */
#include "nt35510.h"
#include "nt35510_reg.h"
#include "app_touchgfx.h"
/* USER CODE END Includes */
```

23. Scroll down to USER CODE BEGIN PV, add the NT35510, and comment out the OTM8009 code:

```
/* USER CODE BEGIN PV */
NT35510_Object_t NT35510Obj;
NT35510_IO_t IOCtx;
//OTM8009A_Object_t OTM8009AObj;
//OTM8009A_IO_t IOCtx;
/* USER CODE END PV */
```

24. In the MX_LTDC_INIT, make the changes to add NT35510 and comment out OTM8009:

```
//OTM8009A_RegisterBusIO(&OTM8009AObj, &IOCtx);
NT35510_RegisterBusIO(&NT35510Obj, &IOCtx);
```

```
//OTM8009A_Init(&OTM8009AObj                ,OTM8009A_FORMAT_RGB888,
OTM8009A_ORIENTATION_LANDSCAPE);
NT35510_Init(&NT35510Obj                    ,NT35510_FORMAT_RGB888,
NT35510_ORIENTATION_LANDSCAPE);
```

```
//HAL_DSI_ShortWrite(&hdsi,      0,    DSI_DCS_SHORT_PKT_WRITE_P1,
OTM8009A_CMD_DISPOFF, 0x00);
HAL_DSI_ShortWrite(&hdsi,        0,    DSI_DCS_SHORT_PKT_WRITE_P1,
NT35510_CMD_DISPOFF, 0x00);
```

25. Save the file.

The change in the control driver is imported. Even though it may seem the older driver is working, eventually the screen would start to get fuzzy as the timing is off. If you were to use the STM32H747I-DISCO LCD driver from the repository, there are further defines that would have to be set to use the correct controller. The need to use a different controller was found while experimenting with the LCD projects found in the STM32Cube repository. Careful reading of the different readme files revealed the issue.

### 15.2.3  Run the Code with the Debugger
Since this is a multicore MCU, we have to set up the debugger like the previous projects.

1. Build the CM7 and CM4 projects. Correct any errors.
2. Right-click on the CM7 project and select Debug AS->STM32 C/C++ Application.
3. The debug configuration dialog appears. We will not halt the cores since the CM4 project doesn't have anything in it but the basic start-up code. Click on the Startup tab.
4. The CM7 project is responsible for downloading and launching both cores. Click on the Add... button.
5. A dialog appears in the Project drop-down. Select STM32H747I-DISCO_CM4.
6. Click OK when finished.

You should see both projects listed; and there is a green arrow next to the CM7 project, indicating that this is the project to start.

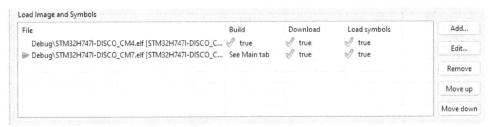

7. Click Apply.
8. Click Ok. The debugger will attempt to start but will fail as the board is not connected yet, but the configuration will be saved.
9. Right-click on the CM4 project and select Debug AS->STM32 C/C++ Application.
10. The debug configuration dialog appears.
11. Click on the Debugger tab.
12. Since the CM7 already has a port and is handling reset, set the following:
    a. Port number: 61238.
    b. Reset behavior: None.
13. Click on the Startup tab.
14. Click on the line in the table for the CM4 project, and click the Edit... button.
15. Uncheck Download. The CM7 debug setup will have already downloaded this code.
16. Click OK when finished.

The debug configuration is now set to NOT download the CM4 project as the CM7 project will download both projects at once.

17. Click Ok.
18. Click Apply.
19. Click Ok. The debugger will attempt to start, but will fail as the board is not connected yet, but the configuration will be saved.
20. Connect the STM32H747I-DISCO board to the development machine.
21. Start the debug session for the CM7 project.
22. The project downloads both CM4 and CM7 projects. The CM7 project runs to a breakpoint in main. Click the continue button.
23. You should see the application appear on the screen. Tap the button a few times and watch the text area message toggle the message.
24. Stop the debugger when finished.

## 15.3 Project 12 Create a GUIX Application

Since ThreadX has a port for Windows, this project will walk through the process of creating a project in Visual Studio that will run on Windows. ThreadX is intended to be run on MCUs that are based on a different architecture than a PC. The ThreadX port for Windows allows developers to learn the features of ThreadX and start application development while hardware is being developed.

This parallel development approach would allow a GUI designer with GUIX Studio to write and test the application on Windows. Once the application is ready, the next step would be to integrate it into the Azure RTOS/MCU platform. GUIX Studio is an application available from the Microsoft Store. With GUIX Studio, you can create the graphic elements of the application, and the project is saved to an XML file with a .gxp extension. When finished, GUIX studio takes the graphic elements and creates C language files that can be integrated into the development tools for the target system. The graphics themselves are turned into pixel hex code.

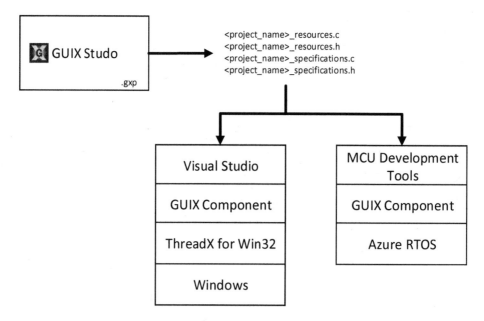

The power of portability allows the output from GUIX Studio to be used on different platforms with different processor architectures. The project for this chapter is to create a simple GUI in GUIX Studio and then use Visual Studio to create a Win32 application using the output from GUIX Studio. The project is broken down into 4 computer activities.

### 15.3.1 Final GUIX Setup Steps
Chapter 3 discusses installing GUIX Studio from the Microsoft Store. The following are the final setup steps:

1. Start GUIX Studio.

2. You will be asked to download the GUIX repository (GUIX component) from GitHub. The interface doesn't work so click the Cancel button.
3. Open a web browser
4. Go to the following address: https://github.com/eclipse-threadx/guix.
5. Click on the Code button and select Download Zip.
6. Save the Zip file to the C:\Azure_RTOS\GUIX-Studio-6.1 folder.
7. Once downloaded, unzip the guix-master.zip file.
8. Rename the unzipped folder from guix-master to guix.

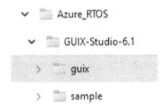

The contents of the quix folder look as follows:

The common and ports folders contain the main source code for the GUIX component. Samples and tutorials provide different examples of how to set up the different widgets or see how a full application is implemented. The samples and tutorials all come with Visual Studio solution files so you can run the sample on the desktop. Finally, graphics and fonts are resource files that are used in the projects. The C:\Azure_RTOS\GUIX-Studio-6.1\guix\ports\win32\build\vs_2019 folder contains the libraries needed for the project.

9.  In the web browser, go to the following address: https://github.com/eclipse-threadx/threadx.
10. Click on the Code button and select Download Zip.
11. Save the Zip file to C:\Azure_RTOS\GUIX-Studio-6.1 folder.
12. Once downloaded, unzip the threadx-master.zip file.
13. Rename the unzipped folder from threadx-master to threadx.

The contents of the threadx folder look as follows:

-Studio-6.1 > threadx

Name

📁 .devcontainer

📁 .pipelines

📁 cmake

📁 common

📁 common_modules

📁 common_smp

📁 docs

📁 ports

📁 ports_arch

📁 ports_module

📁 ports_smp

📁 samples

📁 scripts

📁 test

📁 utility

📄 CMakeLists.txt

📄 CONTRIBUTING.md

📄 LICENSE.txt

📄 LICENSED-HARDWARE.txt

📄 README.md

📄 SECURITY.md

Three folders make up the common ThreadX source files. Four directories that cover different ports to different MCU and MCU architecture types. The samples, test, and utility folders provide demo and diagnostic utilities. The C:\Azure_RTOS\GUIX-Studio-6.1\threadx\ports\win32\vs_2019 folder contains the libraries needed for the project.

### 15.3.2  Create a GUI using GUIX Studio
The simple demo will have a button and a text box. Each time the button is clicked, the text box will toggle between two messages to display.

1. Open File Explore.
2. Create a new folder under C:\Azure_RTOS\GUIX-Studio-6.1 called hello_world

**Note**: never use a dash (-) in a name.

3. Open GUIX Studio.
4. From the menu, select Project->New Project.
5. In the dialog box that appears, enter the following:
   a. Project Name: hello_world
   b. Project Path: C:\Azure_RTOS\GUIX-Studio-6.1\hello_world

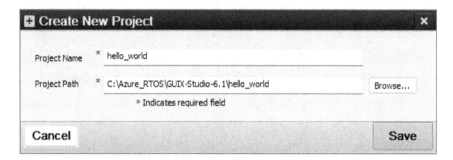

6. Click the Save button.
7. Another dialog appears.
8. Change the following:
   a. X resolution: 800
   b. Y resolution: 0
   c. Name: main_display
9. Click the Save button.

10. The project is created and a dialog appears with a simple help message. Click OK to close the dialog.

If you are familiar with creating Windows applications in Visual Studio, GUIX Studio development is very different. Controls are called Widgets. The main window for the applications is of the Widget Type: window. Each widget has a set of properties that can be configured. There are resources on the right for picture images, fonts, colors, and strings.

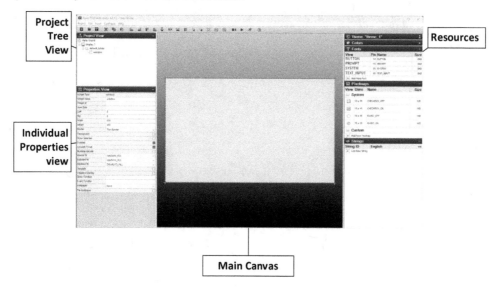

Project Tree View

Resources

Individual Properties view

Main Canvas

11. With the main canvas selected, in the Properties View change Widget Name to hello_world.

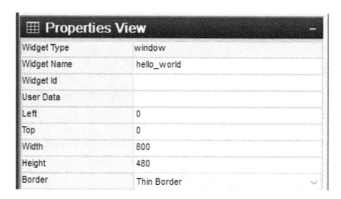

| Properties View | |
|---|---|
| Widget Type | window |
| Widget Name | hello_world |
| Widget Id | |
| User Data | |
| Left | 0 |
| Top | 0 |
| Width | 800 |
| Height | 480 |
| Border | Thin Border |

12. In the Book Exercises Chapter 12 folder, there is a ClickMe.png source that will be used in the project. Expand the Pixelmaps in the resources on the right.
13. Click on Add New Pixelmap.
14. Open the ClickMe.png file. The resource is added to the available resources.

15. Right-click on the canvas.

16. A context menu appears. Select Insert->Button->Pixalmap Button.

17. A button is placed on the canvas. In the properties, set the following:
   a. Widget Id: ID_BUTTON_CLICKME
   b. Width: 141
   c. Height: 40
   d. Normal Pixelmap: CLICKME

18. Use the mouse to move the button to the low center of the canvas.
19. Right-click on the canvas.
20. From the context menu, select Insert->Text->Single Line Input.
21. In the properties set the following:
    a. Width: 300
    b. Height:40
    c. Text: Well, Hello There!
    d. Normal Text Color: Text:
22. Adjust the Single Line Input box to the center, upper half of the canvas.

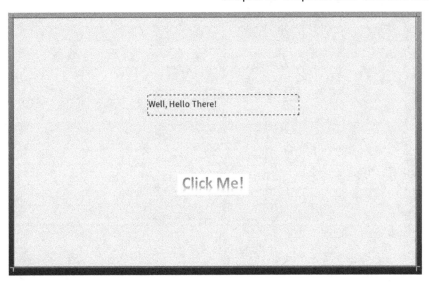

23. Save the project.
24. Now, we need to generate the source code to create the application. From the menu, select Project->Generate All Output Files.
25. A dialog appears asking what to export. Keep the defaults and click the Generate button.

26. A dialog appears when the output files have been generated. Click the OK button.
27. Open File Explorer and go to the C:\Azure_RTOS\GUIX-Studio-6.1\hello-world folder. The C source code files have been generated for the project. We are ready to move to the next step to create an application in Visual Studio.

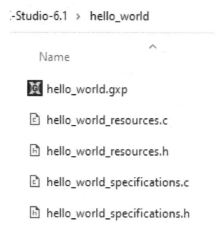

28. Close GUIX Studio.

### 15.3.3 Building the Libraries

The GUIX and ThreadX libraries that are ported to Windows will be linked to our project in Visual Studio. The current libraries have been built with Visual Studio 2019. We will first rebuild the libraries with Visual Studio 2022, so they can easily be linked with the application.

1. Open Visual Studio.
2. Open the guix.sln file under C:\Azure_RTOS\GUIX-Studio-6.1\guix\ports\win32\build\vs_2019.
3. You may be asked to upgrade the solution. Click OK.
4. Leave the build type set to debug x86 (Win32), and build the project.

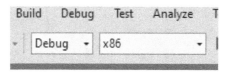

The new gx.lib will be placed under the C:\Azure_RTOS\GUIX-Studio-6.1\guix\ports\win32\build\vs_2019 \Debug_GUIX_5_4_0_COMPATIBILITY folder.

5. Close the project.
6. Open the azure_rtos.sln file under C:\Azure_RTOS\GUIX-Studio-6.1\threadx\ports\win32\vs_2019\example_build. The project is actually building both the library and a demo app. All we will need is the library in the end.
7. You may be asked to upgrade the solution. Click OK.
8. Leave the build type set to debug x86 (Win32) and build the project.

9.  The new tx.lib will be placed under the C:\Azure_RTOS\GUIX-Studio-6.1\threadx\ports\win32\vs_2019\example_build\tx\Debug folder.

Now, we want to update the libraries and needed .h files with the GUIX application.

10.  Create a folder under C:\Azure_RTOS\GUIX-Studio-6.1\ called libraries.
11.  Copy the tx.lib and gw.lib to the C:\Azure_RTOS\GUIX-Studio-6.1\libraries folder.
12.  Copy gx_api.h found under C:\Azure_RTOS\GUIX-Studio-6.1\guix\common\inc to C:\Azure_RTOS\GUIX-Studio-6.1\libraries folder.
13.  Copy tx_api.h found under C:\Azure_RTOS\GUIX-Studio-6.1\threadx\common\inc to the C:\Azure_RTOS\GUIX-Studio-6.1\libraries folder.
14.  Copy tx_port.h found under C:\Azure_RTOS\GUIX-Studio-6.1\threadx\ports\win32\vs_2019\inc to the C:\Azure_RTOS\GUIX-Studio-6.1\libraries folder.
15.  Copy the gx_port.h and gw_win32_display_driver.h found under C:\Azure_RTOS\GUIX-Studio-6.1\guix\ports\win32\inc to the C:\Azure_RTOS\GUIX-Studio-6.1\libraries folder.

The folder should look as follows:

Putting the libraries and necessary .h files in a single location allows them to be re-used for other projects.

### 15.3.4 Creating a Win32 Application

With the libraries created, the final step is to create the application in Visual Studio.

1. Create a new C Application project under C:\Azure_RTOS\GUIX-Studio-6.1\hello_world called helloworld-Test.
2. Set the build type to Debug x86

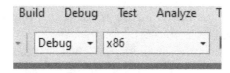

3. We need to import the source files from GUIX Studio. In Solution Explorer, right-click on Source Files and select Add-> Existing item... from the context menu.
4. Locate hello_world_resources.c and hello_world_specification.c and add them to the project.
5. In Solution Explorer, right-click on Header Files and select Add-> Existing item... from the context menu.
6. Locate hello_world_resources.h and hello_world_specification.h and add them to the project.
7. Right-click on Header Files and select Add-> Existing item... from the context menu.
8. Locate the gx_api.h, gx_port.h, gx_win32_display_driver.h, tx_port.h, and tx_api.h and add them to the project.
9. Now we need to configure the properties of the project to point to the library resources. From the menu, select Project>Properties.
10. In the dialog, go down the tree on the left and expand C/C++.
11. Under General, click on the Additional Include Directories.
12. Hit the drop-down and click Edit.
13. A new dialog appears. Click on the new line folder icon and add C:\Azure_RTOS\GUIX-Studio-6.1\libraries to the list.
14. Create another new line and add C:\Azure_RTOS\GUIX-Studio-6.1\hello_world to the list.

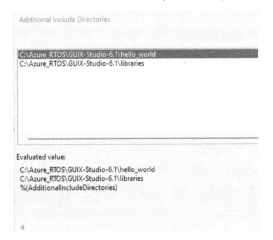

15. Click Ok.
16. Expand the Linker tree.
17. Under General, click on Additional Library Directories.
18. Hit the drop-down and click Edit.
19. A new dialog appears. Click on the new line folder icon and add C:\Azure_RTOS\GUIX-Studio-6.1\libraries to the list.
20. Click OK.
21. Under Input, click on Additional Dependencies.
22. Hit the drop-down and click Edit.
23. A new dialog appears. Enter gx.lib and tx.lib on separate lines.

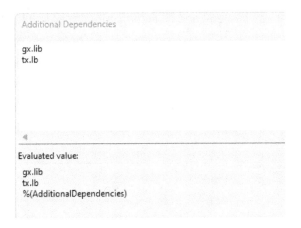

24. Click Ok.
25. Under All Options, change SubSystem to Windows (/SUBSYSTEM:WINDWOS). The change is necessary if this is going to be a Windows application.

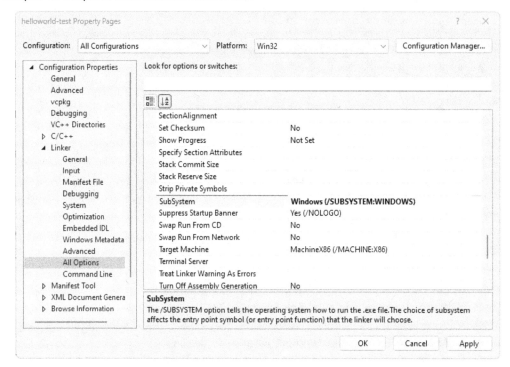

26. Click OK to close the properties dialog.

If you are familiar with developing Windows WPF or Form applications, Visual Studio has a lot of automation. In the Visual Studio Designer, double-clicking on the control will open a source code file that you can edit and fill in the code. For example, button controls have event handlers automatically created and all you have to do is write the code behind the button. GUIX Studio only lets you create the GUI, and you have to add all the code behind the scenes as a later step. There is a little more work to do to create the application. With that understanding, let's write the code, and then we can explain the details.

27. Open Source.c and enter the following:

```c
#include <stdio.h>
#include "tx_api.h"
#include "gx_api.h"

/* Include GUIX resource and specification files for example. */

#include "hello_world_resources.h"
#include "hello_world_specifications.h"

/* Define the new example thread control block and stack. */

TX_THREAD guix_thread;
```

262

```
UCHAR guix_thread_stack[4096];

/* Define the root window pointer. */

GX_WINDOW_ROOT* root_window;

/* Define function prototypes. */

VOID guix_thread_entry(ULONG thread_input);
UINT win32_graphics_driver_setup_565rgb(GX_DISPLAY* display);

int toggle = 0;

int main(int argc, char** argv)
{
    /* Enter the ThreadX kernel. */
    tx_kernel_enter();
    return(0);
}

VOID tx_application_define(void* first_unused_memory)
{
    /* Create the new example thread. */
    tx_thread_create(&guix_thread, "GUIX Thread", guix_thread_entry,
0, guix_thread_stack, sizeof(guix_thread_stack), 1, 1,
TX_NO_TIME_SLICE, TX_AUTO_START);
}

VOID guix_thread_entry(ULONG thread_input)
{

    /* Initialize the GUIX library */
    gx_system_initialize();

    /* Configure the main display. */
    gx_studio_display_configure(MAIN_DISPLAY,   /* Display to
configure*/
        win32_graphics_driver_setup_565rgb, /* Driver to use */
        LANGUAGE_ENGLISH,                       /* Language to install */
        0,          /* Theme to install */
        &root_window);                          /* Root window pointer */

    /* Create the screen - attached to root window. */
    gx_studio_named_widget_create("hello_world",
(GX_WIDGET*)root_window, GX_NULL);

    /* Show the root window to make it visible. */
    gx_widget_show(root_window);

    /* Let GUIX run. */
```

```
    gx_system_start();

}

UINT app_event_handler(GX_WINDOW* window, GX_EVENT* event_ptr)
{

    switch (event_ptr->gx_event_type)
    {

    case GX_SIGNAL(ID_BUTTON_CLICKME, GX_EVENT_CLICKED):

        if (toggle == 0) {
            /* Clear input buffer.  */

gx_single_line_text_input_buffer_clear(&hello_world.hello_world_text_
input);

gx_single_line_text_input_character_insert(&hello_world.hello_world_t
ext_input, "Welcome to", 10);

            toggle = 1;
        }
        else
        {

gx_single_line_text_input_buffer_clear(&hello_world.hello_world_text_
input);

gx_single_line_text_input_character_insert(&hello_world.hello_world_t
ext_input, "Azure RTOS", 10);
            toggle = 0;
        }
        break;

    default:

        return gx_window_event_process(window, event_ptr);
    }

    return 0;
}
```

28. Save the file.

The source code is based on the example source code from the ThreadX documentation. When the application launches, the user can click on the button and the message in the text box will toggle a message. Starting from the top of the source code listing are the header files to include, which consist of the two ThreadX component libraries and the header files from the GUIX Studio output. A thread, thread stack, a GUIX root windows pointer are defined. Function prototypes for the GUIX thread and the display driver are defined. The toggle variable is declared.

The main function makes a call to start the ThreadX kernel, which is part of tx.lib. In the process of starting the kernel, the tx_applicaiton_define function is called, which creates a thread to start the GUIX window and display the GUI we created in GUIX Studio.

The guix_thread_entry is the thread function. A call is made to start the GUIX library, and then the gx_studio_display_configure is called to associate the MAIN_DISPLAY and the video driver that is part of the GUIX library (gx.lib) with the root_window.

In Computer Activity 12.2, we set the display name to "main_display" and the resolution to 800x480 in GUIX Designer. The output from file hello_world_resources.h has defined that information as follows:

```
/* Display and theme definitions
*/

#define MAIN_DISPLAY 0
#define MAIN_DISPLAY_COLOR_FORMAT GX_COLOR_FORMAT_565RGB
#define MAIN_DISPLAY_X_RESOLUTION 800
#define MAIN_DISPLAY_Y_RESOLUTION 480
#define MAIN_DISPLAY_THEME_1 0
#define MAIN_DISPLAY_THEME_TABLE_SIZE 1
```

The MAIN_DSPLAY is used in the gx_studio_display_configure function as the first and only display. The gx.lib source code includes several driver types. The win32_graphics_driver_setup_565rgb was chosen to match the color format supported. The gx_studio_named_widget_create function makes the final connection between the hello_world canvas widget and the root_window. The last two function calls are to show the root_window and start GUIX.

The code has primarily focused on creating and starting the GUI. Events have to be addressed separately. The app_event_handler function handles all the events that take place within the application. There are 63 widget events defined in gx_api.h. When the GUI is launched, the app_event_handler is called. Since there are no matching cases in the switch-case statement, the gx_window_event_process is called as the return. When the ClickMe button is clicked, the app_event_handler is called, the case is matched as a GX_EVENT_CLIEKD, and the text box changes messages based on the toggle value. The button was named ID_BUTTON_CLICKME in GUIX Studio, and the widget id is defined in the GUIX Studio output file hellow_world_specifications:

```
/* Define widget ids
*/

#define ID_BUTTON_CLICKME 1
```

The &hello_world.hello_world_text_input is the address to the text box. The hello_world_specification.h links the widgets to the hello_world application.

```
/* Declare top-level control blocks
*/

typedef struct HELLO_WORLD_CONTROL_BLOCK_STRUCT
{
    GX_WINDOW_MEMBERS_DECLARE
    GX_PIXELMAP_BUTTON hello_world_pixelmap_button;
    GX_SINGLE_LINE_TEXT_INPUT hello_world_text_input;
} HELLO_WORLD_CONTROL_BLOCK;


/* extern statically defined control blocks
*/

#ifndef GUIX_STUDIO_GENERATED_FILE
extern HELLO_WORLD_CONTROL_BLOCK hello_world;
#endif
```

You can see that the names and settings defined in GUIX Studio translate into code and are called in the main application. For the event handler to be called, a link has to be made to the hello_world widget.

29. Open hello_world_specification.h
30. Around line 113 after the comment to defined event process functions, etc. add the following to declare the function:

```
/* Declare event process functions, draw functions, and callback
functions    */
UINT app_event_handler(GX_WINDOW* window, GX_EVENT* event_ptr);
```

31. Save the file.
32. Open hello_world_specification.c. Around line 194 in the GX_CONST GX_STUDIO_WIDGET hello_world_define, replace the GX_NULL with the event function override, app_event_handler:

```
GX_CONST GX_STUDIO_WIDGET hello_world_define =
{
    "hello_world",
    GX_TYPE_WINDOW,                            /* widget type
*/
    GX_ID_NONE,                               /* widget id
*/
    #if defined(GX_WIDGET_USER_DATA)
    0,                                        /* user data
*/
    #endif
```

266

```
    GX_STYLE_BORDER_THIN|GX_STYLE_ENABLED,      /* style flags
*/
    GX_STATUS_ACCEPTS_FOCUS,                     /* status flags
*/
    sizeof(HELLO_WORLD_CONTROL_BLOCK),           /* control block size
*/
    GX_COLOR_ID_WINDOW_FILL,                      /* normal color id
*/
    GX_COLOR_ID_WINDOW_FILL,                      /* selected color id
*/
    GX_COLOR_ID_DISABLED_FILL,                    /* disabled color id
*/
    gx_studio_window_create,                      /* create function
*/
    GX_NULL,                                      /* drawing function
override       */
    (UINT(*)(GX_WIDGET*, GX_EVENT*)) app_event_handler,      /* event
function override        */
    {0, 0, 799, 479},                            /* widget size
*/
    GX_NULL,                                      /* next widget
*/
    &hello_world_pixelmap_button_define,          /* child widget
*/
    0,                                            /* control block
*/
    (void *) &hello_world_properties              /* extended properties
*/
};
```

33. Save the file.
34. Build the application and correct any errors.
35. Start a debug session. The application will start up and show the initial screen:

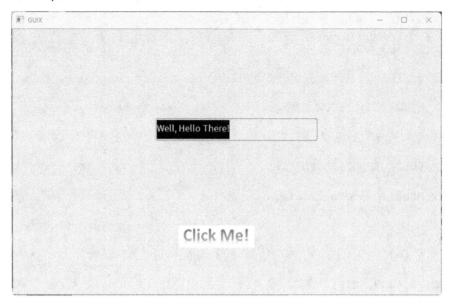

36. Click the ClickMe button a few times to toggle the Welcome to Azure RTOS message.
37. When finished stop the debug session.


## 15.4 GUIX Integration into STM32CubeIDE Project

The next step for this project would be to integrate the application into an STM32CubeIDE project for the STM32H747I-DISCO board. The CM7 sub-project would handle the GUIX application. There were many problems with getting GUIX up and running in this application. Only one example of GUIX running on STM32 was found during the topic research. The developer of this example participates in the STMicroelectronics Community page and was able to answer some questions about his solution. The end project did not run successfully, but there were some lessons learned. Here are the two GUIX projects from the ST Community:

- GitHub - c4chris/H747-Test: STM32 H747I DISCO trying out Azure RTOS ThreadX and GUIX on dual-core and DSI LTDC display - https://github.com/c4chris/H747-Test

- GitHub - c4chris/H747-WeighingStation: Control board for a prototype weighing station, HDMI display and touchscreen USB interface - https://github.com/c4chris/H747-WeighingStation

### 15.4.1 Rebuild the GUIX Application for the Processor

GUIX Studio produces 4 output files, which appear to be pretty simple. Just drop in the 4 files, add the GUIX library, make a few adjustments to project build settings, and everything should work.

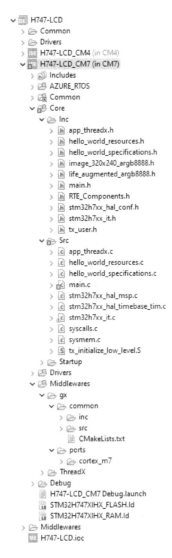

**Note**: Placing the gx folder under the Middlewares folder is not the best organization, since the folder will get wiped out with any changes to the .ioc file.

Except it doesn't work. With no changes, the build will fail. The RAM will be overrun by 270KB.

```
ver/Inc -I../../Drivers/STM32H7xx_HAL_Driver/Inc/Legacy -I../../Drivers/CMSIS/Device/ST/STM32
sections -static --specs=nano.specs -mfpu=fpv5-d16 -mfloat-abi=hard -mthumb -Wl,--start-group
one-eabi/bin/ld.exe: H747-LCD_CM7.elf section `.bss' will not fit in region `RAM_D1'
one-eabi/bin/ld.exe: region `RAM_D1' overflowed by 273544 bytes
```

| Build Analyzer ✕ | Static Stack Analyzer | Cyclomatic Complexity | Call Hierarchy | | | |
|---|---|---|---|---|---|---|
| **H747-LCD_CM7.map** - /H747-LCD_CM7/Debug - Oct 6, 2023, 1:57:35 PM | | | | | | |
| Memory Regions | Memory Details | | | | | |
| Region | Start address | End address | Size | Free | Used | Usage (%) |
| RAM_D1 | 0x24000000 | 0x2407ffff | 512 KB | -272008 B | 777.63 KB | 151.88% |
| FLASH | 0x08000000 | 0x080fffff | 1024 KB | -616948 B | 1.59 MB | 158.84% |
| DTCMRAM | 0x20000000 | 0x2001ffff | 128 KB | 128 KB | 0 B | 0.00% |
| RAM_D2 | 0x30000000 | 0x30047fff | 288 KB | 288 KB | 0 B | 0.00% |
| RAM_D3 | 0x38000000 | 0x3800ffff | 64 KB | 64 KB | 0 B | 0.00% |
| ITCMRAM | 0x00000000 | 0x0000ffff | 64 KB | 64 KB | 0 B | 0.00% |

In GUIX studio, the project settings have to be adjusted to remove the generic allocation for the framebuffer (canvas memory).

1. Open GUIX Studio
2. Open the hello_world project.
3. From the menu select Configure->Project Displays.
4. The Configuration Project dialog appears. Change the following settings:
   a. Target CPU: ST ChromeArt
   b. Toolchain: GNU
   c. Display Configuration: 32 bpp
   d. Uncheck "Allocate canvas memory"

The unchecking of the canvas memory allocation is an important step. You have to create the framebuffer manually for your targeted MCU, since the FMC/SDRAM is in a different memory map location.

5. Click Save.
6. From the menu, select Project->Generate All Output Files.
7. The file can be copied to the STM32CubeIDE project.

### 15.4.2 Additional Changes to the Project
Besides the GUIX configuration settings to target STM32 processors, additional project code steps need to be addressed:

1. A framebuffer needs to be allocated in SDRAM. For the STM32H747I-DISCO board, the location is at 0xD0000000.

**Note**: The STM32H747I-DISCO board has some confusion among the documentation, the schematics, and the sample source files regarding the SDRAM1 location in the memory map. The memory location has been confirmed to be 0xD0000000. The schematic is correct, but all other documents listing the address as 0xC0000000 are wrong.

2. Multiple additions to app_threadx.c file need to be made:
   a. Create a thread to launch the GUIX and the application.
   b. Create the buffer and point to the memory location.
   c. Create code to make changes to the buffer.
   d. Create an events handler.
   e. Create code to handle the touch input. The touch screen is connected to I2C4.

The project builds but crashes in the middle of clearing the SDRAM memory. The project has been made available with the book files if anyone wants to finish it.

### 15.4.3  Interesting LCD Testing Results

While researching the LCD, one of the first steps was to test the STM32H747 repository LCD examples. The STM32F769I-DISCO board was also used in the research testing since it uses the same MB1166-A09 LCD daughter board and was a single-core M7 MCU rather than a dual-core MCU. The LCD_DSI_CmdMode_DoubleBuffer repository example was used on both boards to test the LCD output. The H747 platform testing took a little more work. The H747 driver attempts to do all the configuration and setup and does not use anything from STM32CubeMX with regard to the display override functions. The results are very different between the two boards. The F769 picture came up with a proper screen display:

The H747 picture shows the same screen but shifted horizontally the lines wrapped:

The problem with the H747 is with the driver that was developed for the board. For STM32CubeMX the selections and settings for DMA2D, DSIHOST, and LTDC can be configured. The resulting project pulls in the HAL driver files and the suggested code in MAIN.C. The next step was to manually add the H747 BSP drivers to the project. The build fails since the BSP's LCD driver defines two functions differently than what is defined for the STM32CubeMX setup. The STM32CubeMX code sets up the following:

static void MX_DSIHOST_DSI_Init(void);
static void MX_LTDC_Init(void);

The stm32h747i_discovery_lcd.c defines the same functions as:

__weak HAL_StatusTypeDef MX_DSIHOST_DSI_Init(DSI_HandleTypeDef *hdsi, uint32_t Width, uint32_t Height, uint32_t PixelFormat)
__weak HAL_StatusTypeDef MX_LTDC_Init(LTDC_HandleTypeDef *hltdc, uint32_t Width, uint32_t Height)

Obviously, the build complains about conflicting types. When digging into the stm32h747i_discovery_lcd.c file, it could be seen that the author of the driver was trying to create an all-inclusive driver intended to set up everything with a single call to BSP_LCD_Init (BSP_LCD_Init(uint32_t Instance, uint32_t Orientation)). Any settings in STM32CubeMX do not affect the project. It's as if the driver developer didn't want to create a driver with the flexibility to support STM32CubeMX.

As it turns out, TouchGFX uses the HAL driver and doesn't use stm32h747i_discovery_lcd.c driver at all. The GUIX example that I found also uses the HAL driver and doesn't use stm32h747i_discovery_lcd.c driver.

## 15.5 Summary: The Tale of Two GUI Approaches

The GUI topic took a little more time to research and the results were not as satisfying as I wished they could have been. The business reasons to exclude GUIX in favor of TouchGFX are understood. Both TouchGFX and GUIX have similar approaches to creating a GUI application. During the research for the chapter, it became clear that TouchGFX gets a GUI up and running much faster than trying to integrate GUIX. It is clear that TouchGFX was designed around FreeRTOS and not ThreadX. There is the mixed language use of C and C++ function calls when interacting with TouchGFX and ThreadX. Finally, TouchGFX Designer doesn't generate the latest ioc or project structure that STM32CubeMX or STM32CubeIDE generates, so there is a little inconsistency. The only thing not explored is creating a project in STM32CubeMX, adding the TouchGFX software component, and then integrating the TouchGFX application, which might be the best development path.

GUIX was designed for ThreadX, and the development process is simple as well. A simple file set and GUIX library is all that is needed. There are some nice examples that can be built to run in Windows. The chapter focused on creating a project to run Windows, but integration into the STM32Cube project was way too challenging than it should have been. The lack of repeatable GUIX examples on STM32 is a problem and a concern. If there are no practical MCU specific examples, are developers really using it?

Although not originally planned to be part of the book, hopefully, this chapter has enlightened you on the story of GUI application development for STM32 MCUs.

# 16 MXCHIP® IoT Dev Kit

The MXCHIP IoT Dev Kit was one of the first development boards to demonstrate connecting to Azure IoT Central. The original demonstration ran Arduino; but over time, ThreadX (Azure RTOS) support was made available via the Getting Started Examples. The board features the MXCHIP EWM3166 module, which combines an STM32F412 with a Wireless chip and 16Mbit Flash chip. Since an STM32 MCU is the core of the module, it only makes sense to attempt to create a project with the SMT32Cube tools, since getting ThreadX running on different boards was one of this book's goals. Even though the board might not be currently available, there are some interesting topics worth covering. This chapter is a more to just follow along with, unless you are lucky enough to have the board. The biggest challenge is that there is not much documentation on the board and the EWM3166 module, which adds some black-box investigation to the task.

## 16.1 MXCHIP IoT Dev Kit Overview

Outside of the strange edge connector for power and I/O, the MXCHIP IoT Dev Kit is a very nice little development board. All the basic sensor support for humidity, temperature, pressure, accelerometer, gyroscope, and magnetometer is available. A little 128x64 display, an audio codec, and 2 user buttons round out the board's features. Everything is laid out well and labeled clearly on the board. The concept for the EMW3166 module is interesting. The module addresses the MCU, WiFi with antenna, and storage in a single solution that targets network-connected appliances. The biggest problem with the board and the module is the lack of documentation. The schematic is available, but what is inside the EMW3166 is a mystery. It was obvious that the STM32Cube Programmer tool could be used to interact with the board, and the GSE called out STM32F4xx. What STM32 MCU and what WiFi chip was being used was unknown. MXCHIP was not forth coming with information, which is odd since they want to sell modules. An extensive search yielded the results of what is in the module. The following block diagram is the closest guess to the MXCHIP IoT Dev Kit's actual block diagram.

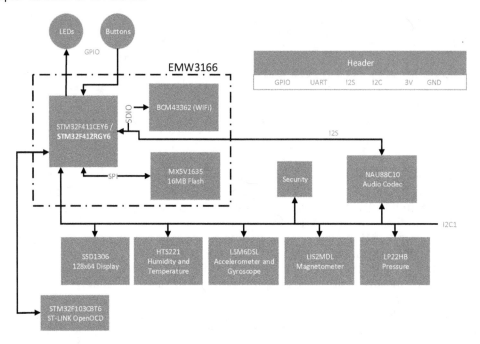

It took some sleuth work to find the exact part number, but the STM32F412RGY6 is the MCU in the EWM3166 module. There is an EWM3165 that uses the STM32F411. Not all the pins get exposed on all package types, so it is very important to get the exact MCU part number. A lesson learned during the investigation when starting this project; assuming what the package was and realizing that a pin was missing for a specific I/O.

## 16.2 Getting Started Example for MXCHIP IoT Dev Kit Review

Now that the exact MCU has been found, one could move forward to develop an STM32Cube project with just the HAL code. Like the STM32L4S5 Discovery Kit, the MXCHIP project structure in the getting started example is a mess. Based on the lack of an .IOC file, it appears the project was created without the aid of the STM32Cube tools; but it looks like the STM32Cube_FW_F4 repository was used. The application is similar to the STM32L4S5 Discovery Kit. Other observations:

- STM32F412RGY6PTR is the MCU part number used in the EWM3166 based on the information gathered.
- A driver is available for the SSD1306 display, and I2C1 is the connection to the MCU.
- Drivers are available for most of the sensors.
- There is no driver for the NAU88C10 Audio Codec or the Security chip.
- MX5V1635 QUAD SPI Flash is only a guess based on the little information that was gathered. There is no flash driver code available.
- The WiFi chip driver is provided as a binary. The actual WiFi chip is old and the original tools to develop the WiFi driver have been deprecated.

276

- The RGB LED is connected to PWMs/TIM.
- 3 LEDs Blue, Yellow, and Green are connected to GPIOs.
- Two user push buttons, A and B, are connected to GPIOs.
- USART6 is used for ST-LINK.
- ST-LINK-OpenOCD is used for debugging rather than the ST-LINK (ST-LINK GBD server).

The example works when built with Visual Code. The challenge is to figure out how to get the project running with the STM32Cube tools. The project is broken into two sub-projects. The first is getting the ThreadX, sensors, LEDs, push buttons, and display enabled and functioning. The second project will integrate the WiFi and NetX Duo support. The getting started example and schematic will be used as a guide for the project.

## 16.3 Project 13 Sensors, LEDs, buttons, and Display: Create the Project with STM32CubeMX

There is no reference design or development kit to base the project on. The project is based on the MCU alone. Between what has been uncovered regarding the hardware and the one available ThreadX (Azure RTOS) project, we can get started on developing this project.

1. Open STM32CubeMX.
2. Under the New Project, click on "Access to MCU Selector".
3. A new window appears. From the Commercial Part Number drop-down, enter STM32F412RGY.
4. On the right-hand side, there is a list of two MCUs associated with the part number. Click on the STM32F412RGY6PTR part number.
5. Click Start Project.

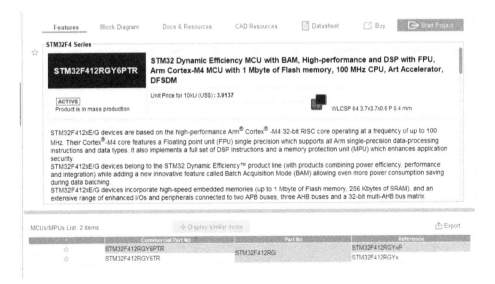

**Note**: The same silicon is packaged in different package types. Not all pins are brought out in all packages. A couple of projects were created with different packages before information was found that led to the correct choice.

### 16.3.1 Enabling the I/O

A major difference with this project is that we are starting from the MCU and not a development kit. With the different IO pins, many things have to be enabled. The first step is to enable the hardware I/O. The board.c source file from the getting started example has some of the information needed to complete the task.

1. Let's start with some of the obvious items. Since I2C1 is connected to all the sensor I/O, we will enable it first. The schematic shows that I2C1 is connected to Pins PB9 and PB8. Under Connectivity, click on ISC1.
2. Set the Mode to I2C.
3. Click on GPIO Settings. The default pin settings for PB6 and PB7 are not correct for this project.

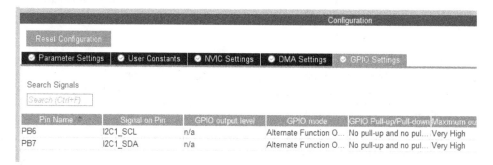

4. Change the I2C mode to disable.
5. From the Pinout view, search for PB9.
6. The pin should flash. Left-click on the pin and select I2C1_SDA.

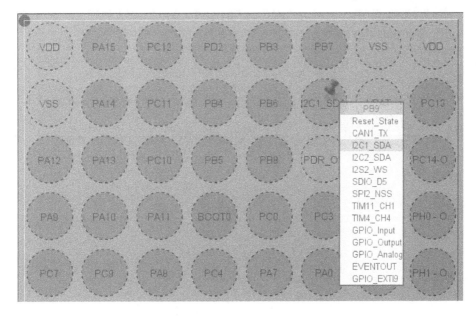

7. Right-click on the PB9 pin and enter the user label: PB9/I2C1_SDA.

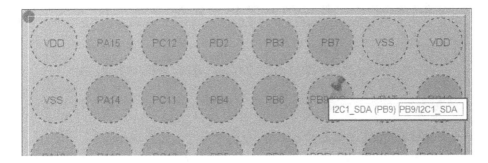

1. From the Pinout view, search for PB8.
2. The pin should flash. Left-click on the pin and select I2C1_SDL.
3. Right-click on the PB8 pin and enter the user label: PB8/I2C1_SDL.
4. Under Connectivity, click on ISC1.
5. Set the Mode to I2C. Now, the I2C pins are set to the correct pints.
6. The board.c file lists how the I2C is set up. In the Configuration section below, click on the Parameter Settings tab and set the following:

    a. I2C Speed Mode: Fast.
    b. Primary Address Length selection: 10-bit.

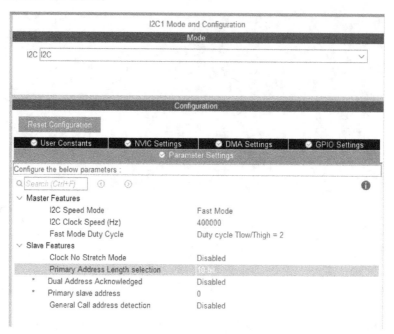

7. For the debug UART port, use USART6. Click on USART6.
8. Set the mode to Asynchronous. The defaults match what is set up in board.c.

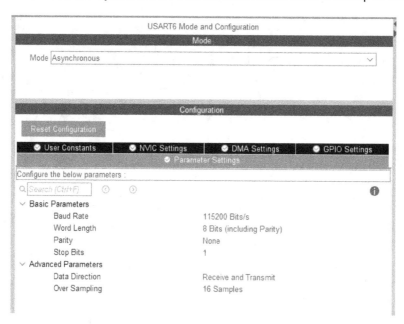

9. The RGP LED is connected to the PWM output from a couple of timers. Expand the Timers section.

10. Per the board.c file and the schematic, TIM2 and TIM3 are connected to the RGB LED. Click on TIM2.
11. Set Channel2 to PWM Generation CH2.
12. Click on GPIO Settings. The PB3 pin matches the schematic. PB3/PWM1 is connected to the RGB Green connector.
13. In the Pinout view on the right, right-click on PB3 and enter the user label: PB3/PWM1.

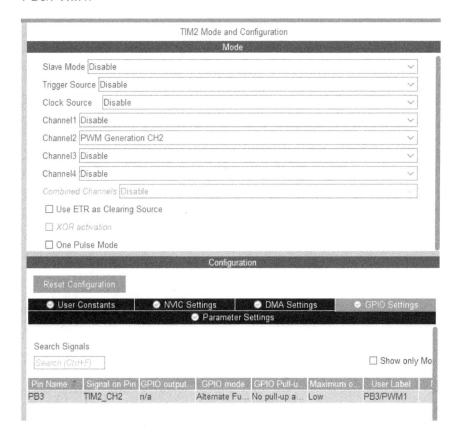

14. Board.c shows the timing for the PWM. Click on the Parameter Setting tab and set the following:

    a. Prescaler: 45.
    b. Counter Period: 2047.

15. Click on TIM3.
16. Set Channel1 to PWM Generation CH1.
17. Locate PB4 in the Pinout view and enter the user label: PB4/PWM2.
18. Set Channel2 to PWM Generation CH2.
19. The GPIO pin that is assigned by default is PB5. The schematic shows PC7. Click on PC7 and select TIM3_CH2. The GPIO listed has been changed.
20. For pin PC7, enter the user label: PC7/PWM3

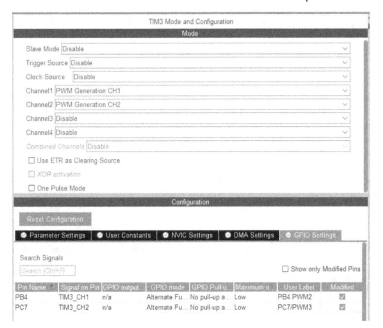

The board_init.c file doesn't need any changes to be made to the PWM parameters, so we will keep the defaults.

21. Now we need to set up the GPIOs for the 3 LEDs and 2 buttons. Looking at the schematic, the A button is connected to PA4. In the Pinout view, set PA4 to GPIO_EXTI4.
22. Set the PA4 user label to PA4/KEYA.
23. The B button is connected to PA10. In the Pintout view, set PA10 to GPIOEXTI10.
24. Set the PA10 user label to PA10/KEYB.
25. The WiFi LED is connected to PB2. In the Pinout view, set PB2 to GPIO_OUTPUT.
26. Set PB2 user label to PB2/LED1
27. The User LED is connected to PC13. In the Pinout view, set PC13 to GPIO_OUTPUT.
28. Set the PC13 user label to PC13/LED2.
29. The Azure LED is connected to PA15. In the Pinout view, set A15 to GPIO_OUTPUT.
30. Set the PA15 user label to PA15/LED3.
31. Expand System Core under Categories.
32. Click on GPIO and then on the GPIO tab. You can check the list of GPIO pins we have enabled.

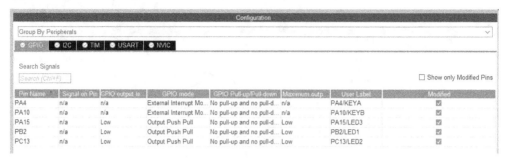

33. Set both the PA4's GPIO Mode and the PA10's GPIO Mode to External Interrupt Mode with Rising/Falling edge trigger detection.

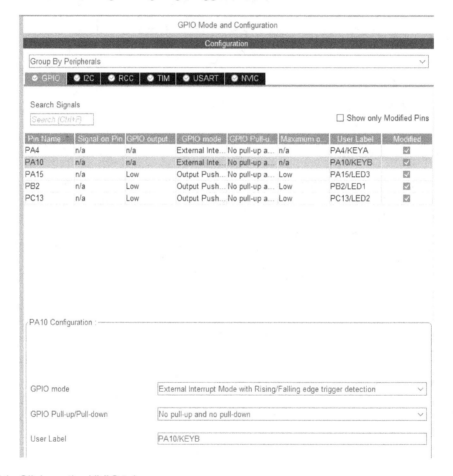

34. Click on the NVIC tab.
35. Enable both GPIO interrupts.

284

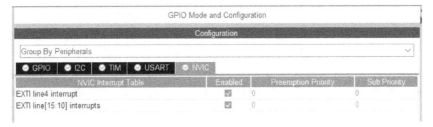

36. Click on RCC under System Core and set the following:
    a. High Speed Clock (HSE): Crystal/Ceramic Resonator.
    b. Low Speed Clock (LSE): Crystal/Ceramic Resonator

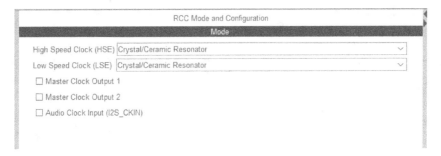

## 16.3.2 Enable Azure RTOS ThreadX Software Package

With the hardware configured, adding the Azure RTOS ThreadX software package is next. Main.c from the getting started example has some information for configuring the Azure RTOS package.

1. Click on the Software Packs drop-down and click on Select Components.
2. Expand STMicroelectronics.X-CUBE-AZRTOS-F4.
3. Expand RTOS ThreadX->ThreadX and check Core.
4. Click Ok. The Select Components dialog closes.
5. Expand Middleware and Software Packs, and click on X-CUBE-AZRTOS-F4.
6. Check the box next to RTOS ThreadX
7. Under the AzureRTOS tab, set the following:
    a. ThreadX memory pool size: 4096

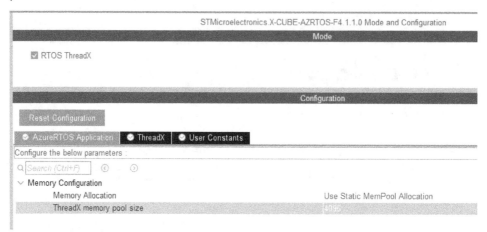

8. Since Azure RTOS has been added, we need to change the SYS clock to one of the timers. Under System Core, click on SYS.
9. Change Timebase Source to TIM7.

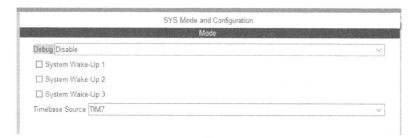

### 16.3.3  Name the Project and Generate the Code

With all the information at hand, that is all that can be configured at this point. If anything was missed, we can always go back to the .IOC file and make changes. Let's save the project and move on to STM32CubeIDE

1. Click on the Project Manager tab.
2. Set the following:

    a. Project Name: MXCHIP_AZ3166.
    b. Project Location: Make the project location the same location as the STM32CubeIDE Workspace folder.
    c. Toolchain/IDE: STM32CubeIDE.
    d. Click on the "Generate Under Root" checkbox.

3. Click on "Generate Code" in the top right. The project will be created in the folder you selected.
4. Once the code has been generated, a dialog will ask you to open the files or open the project. Click on the "Open Project" button.

5. STM32CubeIDE will open and import the project into the workspace folder that you created in Chapter 3.
6. Close STM32CubeMX.


### 16.3.4 Add the Drivers to the Project
The drivers from the getting started example need to be put into the project.

1. Open File Explorer.
2. In the newly created project, created a folder under MXCHIP_AZ3166\Drivers called BSP.
3. Copy the two folders, ssd1306 and stm_sensor, under MXChip\AZ3166\lib\mxchip_bsp to the MXCHIP_AZ3166\Drivers\BSP folder.
4. Close File Explorer.
5. In STM32CubeIDE, refresh the MXCHIP_AZ3166 project. The BSP folder and all the subfolders should appear.

```
∨ 🔲 MXCHIP_AZ3166
  > 🔳 Includes
  > 📂 AZURE_RTOS
  > 📂 Core
  ∨ 📂 Drivers
    ∨ 📂 BSP
      ∨ 📂 Components
        ∨ 📂 ssd1306
          > .h ssd1306_conf.h
          > .c ssd1306_fonts.c
          > .h ssd1306_fonts.h
          > .c ssd1306.c
          > .h ssd1306.h
              📄 LICENSE
        ∨ 📂 stm_sensor
          ∨ 📂 Inc
            > .h hts221_reg.h
            > .h lis2mdl_reg.h
            > .h lps22hb_reg.h
            > .h lsm6dsl_reg.h
            > .h sensor.h
          ∨ 📂 Src
            > .c hts221_read_data_polling.c
            > .c hts221_reg.c
            > .c lis2mdl_read_data_polling.c
            > .c lis2mdl_reg.c
            > .c lps22hb_read_data_polling.c
            > .c lps22hb_reg.c
            > .c lsm6dsl_read_data_polling.c
            > .c lsm6dsl_reg.c
```

6.  The next step is to point the build process to all the include folders. Right-click on the MXCHIP_AZ3166\Drivers\BSP \ssd1306 folder and select "Add/remove include paths…"
7.  A dialog appears, make sure both Debug and Release boxes are checked and click OK.
8.  Right-click on the MXCHIP_AZ3166\Drivers\BSP\stm_sensor\inc folder and select "Add/remove include paths…"
9.  A dialog appears. Make sure both Debug and Release boxes are checked and click OK.
10. Let's double-check the includes. From the menu, select Project->Properties.
11. Go to C/C++ Build->Settings->MCU GCC Compiler->Include paths.

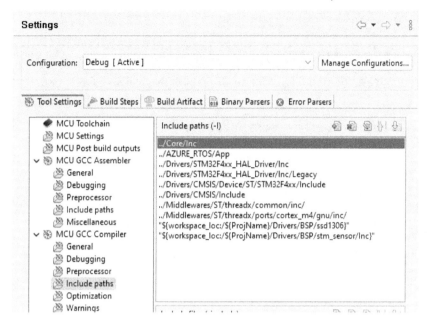

12. Click Apply,

### 16.3.5 Edit the Driver Code

With the project configured, we can edit the code. The application will launch and display some sensor information on the display. The buttons will be used to turn on and off an LED, and an Azure Thread will toggle another LED every second.

1. In STM32CubeIDE open main.c.
2. Scroll down and you will see the definition of the I2C_HandleTypeDef set to hi2cl. The Getting Started Example board_init.c sets the handle to I2cHandle, which is used in the display and sensor drivers. Redefine the handle that STM32CubeMX creates to hi2cl. Open ssd1306.h.
3. Uncomment the define for STM32F4

```
// Choose a microcontroller family
//#define STM32F0
//#define STM32F1
#define STM32F4
//#define STM32L0
//#define STM32L4
//#define STM32F3
//#define STM32H7
//#define STM32F7
```

4. Change the SSD1306_I2C_PORT from I2cHandle to hi2cl.

```
// I2C Configuration
#define SSD1306_I2C_PORT        hi2c1
#define SSD1306_I2C_ADDR        (0x3C << 1)
```

5.  Uncomment the defile for #define SSD1306_INCLUDE_FONT_6x8.

```
// Include only needed fonts
#define SSD1306_INCLUDE_FONT_6x8
// #define SSD1306_INCLUDE_FONT_7x10
#define SSD1306_INCLUDE_FONT_11x18
// #define SSD1306_INCLUDE_FONT_16x26
```

6.  Save and close the file.
7.  Open hts221_read_data_polling.c
8.  Change extern I2C_HandleTypeDef from I2cHandle to hi2c1.
9.  Change #define hi2c1 to #define hi2c1 I2cHandle

```
extern I2C_HandleTypeDef hi2c1;
extern UART_HandleTypeDef UartHandle;

#define hi2c1 I2cHandle
```

10. Replace all three hi2c1 references to I2cHandle.
11. Save and close the file.

12. Open lis2mdl_read_data_polling.c.
13. Change extern I2C_HandleTypeDef from I2cHandle to hi2c1.
14. Change #define hi2c1 I2cHandle to #define hi2c1 I2cHandle.
15. Replace all three hi2c1 references with I2cHandle.
16. Save and close the file.

17. Open lp22hb_read_data_polling.c.
18. Change extern I2C_HandleTypeDef from I2cHandle to hi2c1.
19. Change #define hi2c1 to #define hi2c1 I2cHandle.
20. Replace all three hi2c1 references with I2cHandle.
21. Save and close the file.

22. Open lsm6dsl_read_data_polling.c.
23. Change extern I2C_HandleTypeDef from I2cHandle to hi2c1.
24. Change #define hi2c1 to #define hi2c1 I2cHandle.
25. Replace all three hi2c1 references with I2cHandle.
26. Save and close the file.

### 16.3.6  Final code edits

The drivers are configured to use the right I2C handle. Let's perform the final edits for the application.

1. Open File Explorer.
2. Go to the getting started example \MXChip\AZ3166\app folder.
3. Copy screen.h and paste screen.h into the project's MXCHIP_AZ3166\Core\Inc folder.
4. Go back to the getting started example \MXChip\AZ3166\app folder.
5. Copy screen.c and paste screen.c into the project's MXCHIP_AZ3166\Core\Src folder. The screen.c file has a nice wrapper API for the driver.
6. In STM32CubeIDE, open main.c for the project.
7. Add the following includes:

```
/* USER CODE BEGIN Includes */
#include "sensor.h"
#include <stdio.h>
#include "stm32f4xx_hal.h"
#include "ssd1306.h"
#include "screen.h"
#include <stdlib.h>
#include <string.h>
#include "ssd1306_fonts.h"
/* USER CODE END Includes */
```

8. Add the function prototypes:

```
/* USER CODE BEGIN PFP */
static void Init_MEM1_Sensors(void);
void Init_Screen(void);
/* USER CODE END PFP */
```

9. In the main() function, under USER CODE BEGIN 2 add the following:

```
/* USER CODE BEGIN 2 */
Init_MEM1_Sensors();
Init_Screen();

printf("Hello World\n");

screen_print("Hello!", L0);

lps22hb_t lps22hb_data     = lps22hb_data_read();
float temp = lps22hb_data.temperature_degC;
char tempStr[5];
snprintf(tempStr, 5, "%0.2f", temp);
char tempOutput[15];
strcpy(tempOutput, "Temp:");
strcat(tempOutput, tempStr);
ssd1306_SetCursor(2, L1);
```

```
ssd1306_WriteString(tempOutput, Font_6x8, White);
ssd1306_UpdateScreen();


float press = lps22hb_data.pressure_hPa;
char pressStr[5];
snprintf(pressStr, 5, "%0.2f", press);
char pressOutput[15];
strcpy(pressOutput, "Presr:");
strcat(pressOutput, pressStr);
ssd1306_SetCursor(2, L2);
ssd1306_WriteString(pressOutput, Font_6x8, White);
ssd1306_UpdateScreen();


hts221_data_t hts221_data = hts221_data_read();
float humidity = hts221_data.humidity_perc;
char humidityStr[5];
snprintf(humidityStr,5, "%0.2f", humidity);
char humidOutput[15];
strcpy(humidOutput, "Humid:");
strcat(humidOutput, humidityStr);
ssd1306_SetCursor(2, L3);
ssd1306_WriteString(humidOutput, Font_6x8, White);
ssd1306_UpdateScreen();
/* USER CODE END 2 */
```

10. Scroll down to USER CODE BEGIN 4 and enter the following

```
/* USER CODE BEGIN 4 */
PUTCHAR_PROTOTYPE
{
  /* Place your implementation of fputc here */
  /* e.g. write a character to the USART1 and Loop until the end
of transmission */
  HAL_UART_Transmit(&huart6, (uint8_t *)&ch, 1, 0xFFFF);

  return ch;
}


GETCHAR_PROTOTYPE
{
      uint8_t ch;
      HAL_UART_Receive(&huart6, &ch, 1, HAL_MAX_DELAY);

      /* Echo character back to console */
      HAL_UART_Transmit(&huart6, &ch, 1, HAL_MAX_DELAY);

      /* And cope with Windows */
```

```c
        if (ch == '\r') {
            uint8_t ret = '\n';
            HAL_UART_Transmit(&huart6, &ret, 1, HAL_MAX_DELAY);
        }


    return ch;
}

static void Init_MEM1_Sensors(void)
{
    if (SENSOR_OK != lps22hb_config())
    {
        printf("Init Error Pressure Sensor\r\n");
    }
    if (SENSOR_OK != hts221_config())
    {
        printf("Init Error Humidity-Temperature Sensor\r\n");
    }


}

void Init_Screen(void)
{
    printf("Scanning I2C bus\r\n\t");

    HAL_StatusTypeDef res;
    for (uint16_t i = 0; i < 128; i++)
    {
        res = HAL_I2C_IsDeviceReady(&hi2c1, i << 1, i, 10);
        if (res == HAL_OK)
        {
            char msg[64];
            snprintf(msg, sizeof(msg), "0x%02x", i);
            printf(msg);
        }
        else
        {
            printf(".");
        }
    }
    printf("\r\n\r\n");

    ssd1306_Init();
}

__weak void button_a_callback()
{
```

```
        //USER LED
        //HAL_GPIO_WritePin(PC13_LED2_GPIO_Port,         PC13_LED2_Pin,
GPIO_PIN_SET);

        //Auzre LED
        HAL_GPIO_WritePin(PA15_LED3_GPIO_Port,          PA15_LED3_Pin,
GPIO_PIN_SET);

        //WIFI LED
        //HAL_GPIO_WritePin(PB2_LED1_GPIO_Port,          PB2_LED1_Pin,
GPIO_PIN_SET);

}

__weak void button_b_callback()
{

        //USER LED
        //HAL_GPIO_WritePin(PC13_LED2_GPIO_Port,         PC13_LED2_Pin,
GPIO_PIN_RESET);

        //Auzre LED
        HAL_GPIO_WritePin(PA15_LED3_GPIO_Port,          PA15_LED3_Pin,
GPIO_PIN_RESET);

        //WIFI LED
        //HAL_GPIO_WritePin(PB2_LED1_GPIO_Port,          PB2_LED1_Pin,
GPIO_PIN_RESET);
}


void HAL_GPIO_EXTI_Callback(uint16_t GPIO_Pin)
{
    switch (GPIO_Pin)
    {
        case (PA4_KEYA_Pin):

            button_a_callback();
            break;

        case (PA10_KEYB_Pin):

            button_b_callback();
            break;

        default:
```

```
            break;
    }
}
/* USER CODE END 4 */
```

11. Save the file.
12. Since snprint is being used in main.c, there needs to be an adjustment to the project settings. From the menu, select Project->Properties
13. Go to C/C++ Build->Settings->MCU Settings.
14. Check the box next to "Use float with printf from newlib-nano (-u _printf_float)".
15. Click Apply and Close.

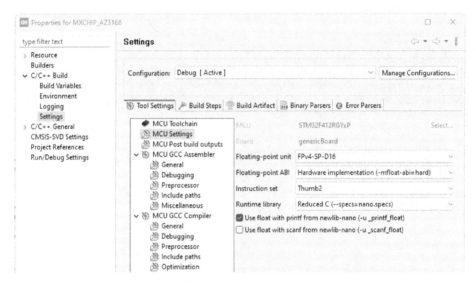

16. Open main.h and add the prototypes for printf over UART:

```
/* USER CODE BEGIN Private defines */
#define PUTCHAR_PROTOTYPE int __io_putchar(int ch)
#define GETCHAR_PROTOTYPE int __io_getchar(void)
/* USER CODE END Private defines */
```

17. Save and close the file.
18. Open app_threadx.c file. A thread will be created to toggle the user LED (yellow).
19. Add the following include:

```
/* USER CODE BEGIN Includes */
#include "main.h"
#include <stdio.h>
/* USER CODE END Includes */
```

20. Set the thread stack size to 1024

```
/* USER CODE BEGIN PD */
#define THREAD_STACK_SIZE 1024
/* USER CODE END PD */
```

21. Add the variables

```
/* USER CODE BEGIN PV */
uint8_t LED_thread_stack[THREAD_STACK_SIZE];
TX_THREAD led_thread_ptr;
/* USER CODE END PV */
```

22. Add the code to create the thread:

```
/* USER CODE BEGIN App_ThreadX_Init */
(void)byte_pool;
tx_thread_create(&led_thread_ptr,                    "led_thread",
my_threledtoggle_thread_entryad_entry, 0x1234, LED_thread_stack,
THREAD_STACK_SIZE, 15,15,1,TX_AUTO_START);
/* USER CODE END App_ThreadX_Init */
```

23. Add the thread function:

```
/* USER CODE BEGIN 1 */
VOID ledtoggle_thread_entry(ULONG initial_input){
        printf("LED Thread Entry Reached\n");
        printf("\n");


        printf("Starting Loop...\n");
        while(1){
            HAL_GPIO_TogglePin(PC13_LED2_GPIO_Port,
PC13_LED2_Pin); /*Defined in main.h*/
            tx_thread_sleep(500);
        }
}
/* USER CODE END 1 */
```

24. Save and close the file.


### 16.3.7 Build and Debug with ST-LINK OpenOCD
Let's test the code completed so far.

1. Build the project and correct any errors.
2. Start the debug session. The debug configuration appears.
3. Click on the Debugger tab.
4. Set the Debug probe to ST-LINK (OpenOCD).

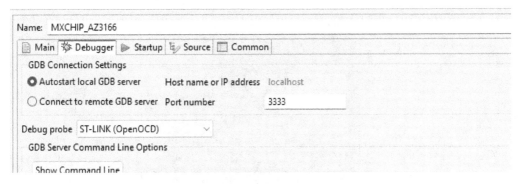

5. Click on the "Show generator options…".
6. Change the Frequency to 4MHz. If left in the default of 8MHz the debug output will display that it makes an attempt to connect at 8MHz; and upon failing, then drops to 4MHZ. By making the change to 4MHZ here, the debugger makes the connection faster.

7. Click Apply and then Click OK. The debugger starts and hits the HAL_Init() breakpoint in main().
8. Click continue. The display will show the temperature, pressure, and humidity readings with a Hello! String on top. The User LED will flash on and off. Press the A and B buttons to toggle the Azure LED (blue).

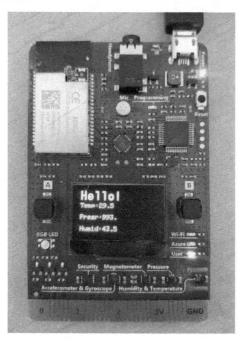

9.  Stop debugging when finished.

## 16.4 Project 14: Clock Configuration Changes

The Getting Started Example adjusts the clock close to the maximum frequency of 100MHz. Let's make a similar adjustment.

1.  In STM32CubeIDE, open the MXCHIP_AZ3166.ioc file.
2.  Click on the Clock Configuration tab. The clock diagram appears. The default MCU clock speed is 16MHz.

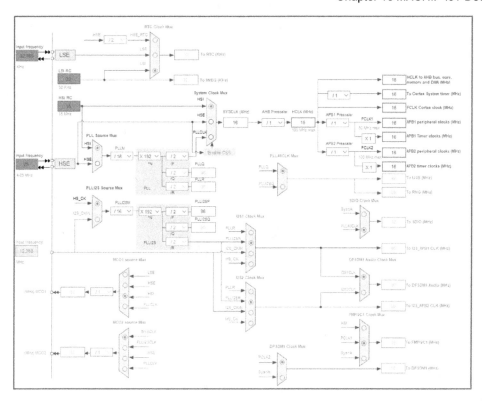

3. Per board_init.c, we will adjust the timing as follows:
   a. Set the PLL Source MUX to HSE.
   b. Change the PLLM to /13.
   c. In the blue PLL section, change *N (PLLN) to X 96.
   d. Set the System Clock Mux to PLLCLK.
   e. Set APB1 Prescaler to /2.
   f. Set PLLI2SM to /13.
   g. Set PLLI2SM *N to x 96.

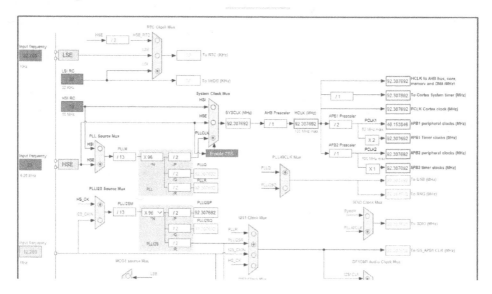

4. Save changes.
5. Rebuild the project.
6. Run the debugger and verify that the code runs as before.


## 16.5 Project 15: Add WiFi and NetX Duo

Now, we can add the WiFi support. The WiFi is connected to SDIO, which is not enabled in Board_init.c; so it must be enabled in the provided WiFi binary.


### 16.5.1 Enable Azure RTOS NetX Duo Software Package
With the hardware configured for the moment, adding the Azure RTOS software packages is next. Main.c and wwwd_networking.c from the getting started example has some information for configuring Azure RTOS packages.

1. In STM32CubeIDE, open the MXCHIP_AZ3166.ioc file.
2. Click on the Software Packs drop-down and click on Select Components.
3. Expand STMicroelectronics.X-CUBE-AZRTOS-F4.
4. Expand Network NetXDuo->NetXDuo and check the following:
    a. NX Core
    b. Addons DHCP Client

Since we only want to get the network running and be able to ping the board remotely, we will only choose these components for now:

5. Click Ok. The Select Components dialog closes.
6. Expand Middleware and Software Packs and click on X-CUBE-AZRTOS-F4/.
7. Check the box next to Network NetXDuo.
8. Under the AzureRTOS tab, set the following:

    c.   NetXDuo memory pool size: 2048.

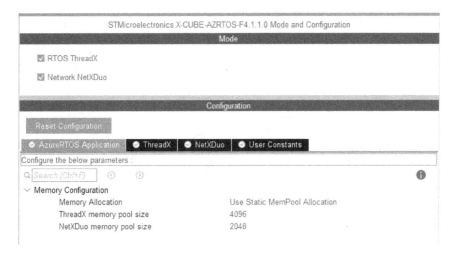

## 16.5.2 Add the Driver to the Project

Unlike the device drivers that come with source code, the WiFi driver comes as a binary library file.

1. Open File Explorer.
2. In the MXChip\AZ3166\lib\ folder, copy the wiced_sdk folder to the MXCHIP_AZ3166\Drivers\BSP folder.
3. Delete the binary_build sub folder in MXCHIP_AZ3166\Drivers\BSP\wiced_sdk.
4. Close File Explorer.
5. In STM32CubeIDE, right-click on the MXCHIP_AZ3166 project and select Refresh. The driver should appear.

6. Right-click on the MXCHIP_AZ3166\Drivers\BSP\wiced_sdk\inc and select "Add/remove include paths..."
7. A dialog appears, make sure both Debug and Release boxes are checked and click OK.
8. The WiFi libwiced_sdk_bin.a needs to be added as a library source. From the menu, select Project->Properties.
9. Go to C/C++ Build->Settings->MCU GCC Linker->Libraries.
10. The name of the library and the path to the library need to be set. In the Libraries box, click on the new icon.
11. Enter wiced_sdk_bin. The "lib" prefix and ".a" are not needed for this driver.
12. In the Libraries search path box, click on the new icon.
13. Click the Workspace button and open \Drivers\BSP\Components\wiced_sdk/lib.
14. Click OK.

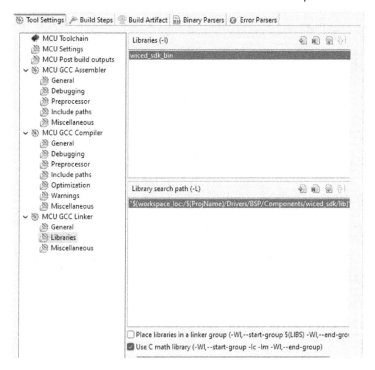

15. Click Apply Close.

### 16.5.3 Add Source Files from the Getting Started Example
We will re-use and then modify the source files from the getting started example.

1. Open File Explorer.
2. In the getting started example's MXChip\AZ3166\app folder, copy azure_config.h, azure_device_x509_cert_config.h, azure_pnp_info.h, and wwd_networking.h to the MXCHIP_AZ3166\NetXDuo\App folder.
3. In the getting started example's MXChip\AZ3166\app folder, copy wwd_networking.c to the MXCHIP_AZ3166\NetXDuo\App folder.
4. In the getting started example's \MXChip\AZ3166\lib\netxduo folder copy nx_user.h file to the MXCHIP_AZ3166\NetXDuo\App folder and overwrite the current file.

### 16.5.4 Edit the code
Since STM32CubeMX sets up the project to run the NetXDuo with the app_netxduo.c file, all we need to do is modify this file to enable the WiFi.

1. In STM32CubeIDE, open app_netxduo.c
2. Add the following includes:

```
/* USER CODE BEGIN Includes */
#include <stdio.h>
#include "tx_api.h"
#include "screen.h"
#include "wwd_networking.h"
#include "azure_config.h"
/* USER CODE END Includes */
```

3. In the MX_NetXDuo_Init() function, add the following:

```
/* USER CODE BEGIN MX_NetXDuo_Init */

UINT status;

if   ((status   =   wwd_network_init(WIFI_SSID,   WIFI_PASSWORD,
WIFI_MODE)))
{
   printf("ERROR: Failed to initialize the network (0x%08x)\r\n",
status);
 }
if((status = wwd_network_connect()))
{
      printf("ERROR:   Failed   to   initialize   the   network
(0x%08x)\r\n", status);
}
/* USER CODE END MX_NetXDuo_Init */
```

4. Save the file.
5. Modify Azure_config.h to include your WiFi router settings.
6. Save the file.


### 16.5.5  Build and Debug the WiFi Additions
The code setup was simple, but the results are not going to work.

1. Open wwd_networking.c.
2. Set a breakpoint at line 61, which has a printf statement stating: "Initializing WiFi".
3. Run the debugger.
4. Once the breakpoint is hit in main, click continue.
5. The breakpoint in wwd_networking.c is reached.
6. Step through the code, and you will get an error with the call to (wwd_management_wifi_on(WIFI_COUNTRY).

```
75
76      // Set country
77      if (wwd_management_wifi_on(WIFI_COUNTRY) != WWD_SUCCESS)
78      {
79          printf("ERROR: wwd_management_wifi_on\r\n");
80          return NX_NOT_SUCCESSFUL;
81      }
82
```

7.  The WiFi setup fails. Stop debugging

### 16.5.6  WiFi Investigation

Without the driver source code or any documentation, it is impossible to determine why the WiFi is failing. In the white paper "Azure RTOS and the MXCHIP IoT DevKit", we demonstrated the WiFi was working successfully when building the getting started example with VScode. Digging into the getting started example project didn't yield much information.

- MMC HAL driver – Adding the MMC HAL driver didn't resolve anything. The getting started example only includes the HAL drivers that were already added to the project. Check the CMakelist.txt in \getting-started\MXChip\AZ3166\lib\stm32cubef4:

find_package(STM32HAL REQUIRED COMPONENTS dma flash gpio tim timebase_tim uart usart i2c)

- Adding HAL usart doesn't resolve the issue.
- The linker script (.ld) actually breaks the code so the sensors drivers fail. The default linker script works just fine.
- Enabling SDIO with MMC 4bit crashes the build as the SDIO driver functions conflict with the same driver functions in the wiced_sdk_bin.a binary file.

There might be something very subtle that needs to be changed in the project to get WiFi working. Considering that this board is available in limited supply, the WiFi support will have to be left incomplete.

## 16.6  Project 16: Back out WiFi Changes and Modify Code a Little More

Projects 13 and 14 used the sensors to get a simple reading. There are more sensors on the board. In this project, the buttons will be used to cycle through all the sensors' output and display data on the display.

### 16.6.1 Back out the WiFi Changes
Let's remove the WiFi and NetX Duo support from the project.

1. Delete the following files from the MXCHIP_AZ3166\NetXDuo\App folder:

    - azure_config.h
    - azure_device_x509_cert_config.h
    - azure_pnp_info.h
    - wwd_networking.h
    - wwd_networking.c
    - nx_user.h

2. From the menu, select Project->Properties.
3. Go to C/C++ Build->Settings->MCU GCC Linker->Libraries.
4. Delete the entries for the wiced_sdk_bin file.
5. Click the "Apply and Close" button.
6. Open the MXCHIP_AZ3166.ioc file.
7. Expand the Middleware and Software Packs.
8. Click on X-CUBE-AZRTOS-F4.
9. Uncheck "NetworkNetXDuo".
10. Save and close the MXCHIP_AZ3166.ioc file.
11. Delete the netxduo folder under the MXCHIP_AZ3166\Middlewares\ST folder.
12. Delete the wiced_sdk folder under the MXCHIP_AZ3166\Drivers\BSP folder.
13. Clean the project.
14. Build the project to make sure there are no errors.

### 16.6.2 Modify the Code
With the WiFi support removed, we can move on to editing the code.

1. In STM32CubeIDE, open ssd1306_conf.h.
2. Uncomment the line for the SSD1306_INCLUDE_FONT_7x10.

```
// Include only needed fonts
#define SSD1306_INCLUDE_FONT_6x8
#define SSD1306_INCLUDE_FONT_7x10
#define SSD1306_INCLUDE_FONT_11x18
// #define SSD1306_INCLUDE_FONT_16x26
```

3. Save and close the file.
4. Open main.h.
5. Add the following extern global variable to the USER CODE section:

```
/* USER CODE BEGIN EM */
extern int sensor_cycle;
/* USER CODE END EM */
```

6. Save and close the file.
7. Open main.c.
8. In Init_MEM1_Sensors, add the initialization of the remaining two sensors:

```
static void Init_MEM1_Sensors(void)
{
    if (SENSOR_OK != lps22hb_config())
    {
        printf("Init Error Pressure Sensor\r\n");
    }
    if (SENSOR_OK != hts221_config())
    {
        printf("Init Error Humidity-Temperature Sensor\r\n");
    }
    if (SENSOR_OK != lsm6dsl_config())
    {
        printf("Init Error Accelerometer Sensor\r\n");
    }
    if (SENSOR_OK != lis2mdl_config())
    {
        printf("Init Error Magnetometer Sensor\r\n");
    }

}
```

9. In button_a_callback(), add the following code after the LED WritePin calls:

```
__weak void button_a_callback()
{

    //USER LED
    //HAL_GPIO_WritePin(PC13_LED2_GPIO_Port,    PC13_LED2_Pin,
GPIO_PIN_SET);

    //Auzre LED
    HAL_GPIO_WritePin(PA15_LED3_GPIO_Port,    PA15_LED3_Pin,
GPIO_PIN_SET);

    //WIFI LED
    //HAL_GPIO_WritePin(PB2_LED1_GPIO_Port,    PB2_LED1_Pin,
GPIO_PIN_SET);

    if (sensor_cycle == 0 ){
        sensor_cycle = 1;
    }
    else if (sensor_cycle == 1){
        sensor_cycle = 2;
```

```
        }
    else if (sensor_cycle == 2){
        sensor_cycle = 0;
    }

}
```

The code changes the value of sensor_cycle on each press of button A. The sensor value is what is displayed on the screen.

10. In button_a_callback(), add the following code after the LED WritePin calls:

```
__weak void button_b_callback()
{

    //USER LED
    //HAL_GPIO_WritePin(PC13_LED2_GPIO_Port,    PC13_LED2_Pin,
GPIO_PIN_RESET);

    //Auzre LED
    HAL_GPIO_WritePin(PA15_LED3_GPIO_Port,    PA15_LED3_Pin,
GPIO_PIN_RESET);

    //WIFI LED
    //HAL_GPIO_WritePin(PB2_LED1_GPIO_Port,    PB2_LED1_Pin,
GPIO_PIN_RESET);

    if (sensor_cycle == 0 ){
        sensor_cycle = 2;
    }
    else if (sensor_cycle == 1){
        sensor_cycle = 0;
    }
    else if(sensor_cycle == 2){
        sensor_cycle = 1;
    }

}
```

The code is the reverse of button A and cycles through what will be displayed in reverse order.

11. Save and close the file.
12. Open app_threadx.c.
13. Add the following includes:

```
/* USER CODE BEGIN Includes */
#include "main.h"
#include <stdio.h>

#include "sensor.h"
#include "stm32f4xx_hal.h"
#include "ssd1306.h"
#include "screen.h"
#include <stdlib.h>
#include <string.h>
#include "ssd1306_fonts.h"
/* USER CODE END Includes */
```

14. Add the variables for a new sensor thread and the sensor_cycle variable.

```
/* USER CODE BEGIN PV */
uint8_t LED_thread_stack[THREAD_STACK_SIZE];
TX_THREAD led_thread_ptr;

uint8_t sensors_thread_stack[THREAD_STACK_SIZE];
TX_THREAD sensors_thread_ptr;

int sensor_cycle = 0;
/* USER CODE END PV */
```

15. Add the prototype for the new sensor thread:

```
/* USER CODE BEGIN PFP */
VOID ledtoggle_thread_entry(ULONG intial_input);

VOID sensors_thread_entry(ULONG intial_input);
/* USER CODE END PFP */
```

16. In App_ThreadX_Init, create the new thread:

```
UINT App_ThreadX_Init(VOID *memory_ptr)
{
  UINT ret = TX_SUCCESS;
  TX_BYTE_POOL *byte_pool = (TX_BYTE_POOL*)memory_ptr;

  /* USER CODE BEGIN App_ThreadX_Init */
  (void)byte_pool;
  tx_thread_create(&led_thread_ptr,                     "led_thread",
ledtoggle_thread_entry,          0x1234,          LED_thread_stack,
THREAD_STACK_SIZE, 15,15,1,TX_AUTO_START);
```

```
    tx_thread_create(&sensors_thread_ptr,           "senors_thread",
sensors_thread_entry,          0x2345,          sensors_thread_stack,
THREAD_STACK_SIZE, 15,15,1,TX_AUTO_START);
    /* USER CODE END App_ThreadX_Init */


    return ret;
}
```

17. Finally, add the new sensor_thread_entry() function after the ledtoggle_thread_entry() function:

```
VOID sensors_thread_entry(ULONG intial_input){

        lps22hb_t lps22hb_data;
        hts221_data_t hts221_data;
        lsm6dsl_data_t lsm6dsl_data;
        lis2mdl_data_t lis2mdl_data;


        while(1){

                if(sensor_cycle == 0){

                        ssd1306_Fill(Black);
                        ssd1306_SetCursor(2, L0);
                        ssd1306_WriteString("Barometer",      Font_7x10,
White);

                        lps22hb_data      = lps22hb_data_read();
                        float temp = lps22hb_data.temperature_degC;
                        char tempStr[5];
                        snprintf(tempStr, 5, "%0.2f", temp);
                        char tempOutput[15];
                        strcpy(tempOutput, "Temp:");
                        strcat(tempOutput, tempStr);
                        ssd1306_SetCursor(2, L1);
                        ssd1306_WriteString(tempOutput,       Font_6x8,
White);

                        float press = lps22hb_data.pressure_hPa;
                        char pressStr[5];
                        snprintf(pressStr, 5, "%0.2f", press);
                        char pressOutput[15];
                        strcpy(pressOutput, "Presr:");
                        strcat(pressOutput, pressStr);
                        ssd1306_SetCursor(2, L2);
                        ssd1306_WriteString(pressOutput,       Font_6x8,
White);
```

```
                        hts221_data = hts221_data_read();
                        float humidity = hts221_data.humidity_perc;
                        char humidityStr[5];
                        snprintf(humidityStr,5, "%0.2f", humidity);
                        char humidOutput[15];
                        strcpy(humidOutput, "Humid:");
                        strcat(humidOutput, humidityStr);
                        ssd1306_SetCursor(2, L3);
                        ssd1306_WriteString(humidOutput,      Font_6x8,
White);


                        ssd1306_UpdateScreen();


            }
            if(sensor_cycle == 1){

                        ssd1306_Fill(Black);
                        ssd1306_SetCursor(2, L0);
                        ssd1306_WriteString("Accelerometer",
Font_7x10, White);

                        lsm6dsl_data = lsm6dsl_data_read();
                        float xa = lsm6dsl_data.acceleration_mg[0];
                        float ya = lsm6dsl_data.acceleration_mg[1];
                        float za = lsm6dsl_data.acceleration_mg[2];

                        char xaStr[5];
                        snprintf(xaStr, 5, "%0.2f", xa);
                        char xaOutput[15];
                        strcpy(xaOutput, "X:");
                        strcat(xaOutput, xaStr);
                        ssd1306_SetCursor(2, L1);
                        ssd1306_WriteString(xaOutput,      Font_6x8,
White);

                        char yaStr[5];
                        snprintf(yaStr, 5, "%0.2f", ya);
                        char yaOutput[15];
                        strcpy(yaOutput, "Y:");
                        strcat(yaOutput, yaStr);
                        ssd1306_SetCursor(2, L2);
                        ssd1306_WriteString(yaOutput,      Font_6x8,
White);

                        char zaStr[5];
                        snprintf(zaStr, 5, "%0.2f", za);
```

311

```c
            char zaOutput[15];
            strcpy(zaOutput, "Z:");
            strcat(zaOutput, zaStr);
            ssd1306_SetCursor(2, L3);
            ssd1306_WriteString(zaOutput,           Font_6x8,
White);


            ssd1306_UpdateScreen();

        }
        if (sensor_cycle == 2){
            ssd1306_Fill(Black);
            ssd1306_SetCursor(2, L0);
            ssd1306_WriteString("Magnetometer",
Font_7x10, White);


            lis2mdl_data = lis2mdl_data_read();

            float xm = lis2mdl_data.magnetic_mG[0];
            float ym = lis2mdl_data.magnetic_mG[1];
            float zm = lis2mdl_data.magnetic_mG[2];


            char xmStr[5];
            snprintf(xmStr, 5, "%0.2f", xm);
            char xmOutput[15];
            strcpy(xmOutput, "X:");
            strcat(xmOutput, xmStr);
            ssd1306_SetCursor(2, L1);
            ssd1306_WriteString(xmOutput,           Font_6x8,
White);


            char ymStr[5];
            snprintf(ymStr, 5, "%0.2f", ym);
            char ymOutput[15];
            strcpy(ymOutput, "Y:");
            strcat(ymOutput, ymStr);
            ssd1306_SetCursor(2, L2);
            ssd1306_WriteString(ymOutput,           Font_6x8,
White);


            char zmStr[5];
            snprintf(zmStr, 5, "%0.2f", zm);
            char zmOutput[15];
            strcpy(zmOutput, "Z:");
            strcat(zmOutput, zmStr);
            ssd1306_SetCursor(2, L3);
```

```
                    ssd1306_WriteString(zmOutput,           Font_6x8,
White);

                    ssd1306_UpdateScreen();
        }



            tx_thread_sleep(100);
        }

    }
```

Two sensors provide the Barometer output. The other two sensors provide accelerometer and magnetometer output respectively. What is displayed on the screen depends on the sensor_cycle value.

18. Save and close the file.

### 16.6.3  Build and Debug the Updated Project

With the code additions complete, let's build and run the project on the board.

1. Build the project and correct any errors.
2. Make sure the board is connected to the development system via the USB cable,
3. Start the debugger.
4. Once the breakpoint is hit, click continue. The display should show the barometer output.
5. Press button A and the accelerometer output is displayed.
6. Press button A again and the magnetometer output is displayed.
7. Press buttons A and B to cycle back and forth through the different screens.
8. Stop debugging when finished.

## 16.7 Summary

Although we couldn't get WiFi working, many little things were covered in this chapter's four projects that round out everything introduced in the book. Rather than starting with a development kit, project 13 created the project from an MCU, which is probably where most developers would start with custom designs. The adjustment of the I2C pin is not obvious as the tools start with a default pin assignment. The second project demonstrated how to adjust the clock settings to meet the maximum frequency of the MCU. The WiFi project covered how to add a binary library file. The final project exercises the remaining sensors that were supported in the Getting Started example project.

# 17 Final Review

As discussed in the preface, the book started as a white paper and turned into something more comprehensive. The two guiding questions helped to propel broader research into the STM32Cube tools and ThreadX than was originally planned for the white paper. With all the efforts to address the two main questions, some areas have been left open such as USBX, wired connection to Azure IoT HUB, connection to the new Azure Event Grid, sensor data from M4 core displayed in TouchGFX application running on M7 core, getting GUIX up and running, and WiFi support for the MXCHIP.

## 17.1 Support and the ST Community

Many problems slowed down the research and the writing of the book. There are many little things such as breaking apart the GSE, adding the library binary, the debugging of two cores, and GUIX setting options that were not widely covered or not clearly explained that took some effort to resolve. The support from the ST Community, online resources from STMicroelectronics, and general web searches were able to resolve many problems. There is no direct support from Microsoft. The support forum on Microsoft Learn for ThreadX (Azure RTOS) is, basically, a handoff to the ST Community for help. Now that ThreadX is part of the Eclipse Foundation, how support will be handled with the Eclipse Foundation is not known.

## 17.2 Summary of STM32Cube and STM32 Development Hardware

Other MCUs and toolsets were considered to see what might be the best path forward to address the two driving questions. STMicroelectronics made the investment to include ThreadX (Azure RTOS) into STM32Cube tools, thus it was an easy choice to continue that research.

As it turns out, STMicroelectronics has done a fantastic job with the STM32Cube Tools and development hardware is affordable. Developers can start developing with low upfront cost. The real cost is learning all the little nuances of the software provided.

Keep in mind that not everything is perfect. There is work that still has to be done. Here is what I found during my journey. The tools were updated several times, which is a good thing to know. It indicates that ST is continuing to invest in the development tools. The tools themselves need a little more alignment, for example, TouchGFX creates a very different STM32CubeIDE project than STM32CubeMX. There are many differences among MCU families when creating a project. The new MCUs have more features enabled to make

315

development easier. For example, the STM32H747 provided a quick way to add a thread to the ThreadX (Azure RTOS) application that was not present for the other MCUs. The newer STM32U5 Discovery Kit didn't require going into Software Packs to select ThreadX components, which is very different from all the other platforms that were covered in the book, and I liked that I didn't have to go those extra steps. The bottom line is that the newer STM32 MCU and development boards are going to have improved support in the tools over the older STM32 MCU and development boards. The QSPI flash setup for the STM32L4S5 Discovery board was incorrect, thus a different platform was needed to demonstrate FileX on non-volatile memory. The repository examples for each MCU are not up to date and required some work to implement. Launching a repository example .ioc file requires an update that might not fit all the previous options or software packages that were available in older tool versions. Also, there is a little lack of clarity regarding what the device drivers provided in the MCU repositories support or what they represent. Since the addition of ThreadX (Azure RTOS) is relatively new, there are very few ThreadX (Azure RTOS) repository examples. One feature that would be nice to see added is a visual code map. The current tools do not make it easy to determine what code modules a particular function is being called from. A visual map can show where a function is being called in the various source files. The missing NetX Duo sub-components should be added. The ST Community is a real plus. The responses are not perfect, but they get you going in the right direction.

The biggest piece missing is the Azure IoT C SDK. Azure is in the name Azure RTOS for a reason so why is the SDK missing? With the goal to make it easy to develop with STM32, it would be nice to not have to struggle to integrate the Azure IoT C SDK into the project. There is an obvious reason for this, which will be discussed in the next section.

The investment into the STM32Cube tools and ThreadX (Azure RTOS) software packages was a wise move by STMicroelectronics. Any developer should be able to get ThreadX up and running on any STM32 MCU fairly quickly. In addition, STMicroelectronics has created a development kit or reference design for every single STM32 MCU available. Having the right tools and development hardware provides a great starting point for design.

## 17.3 Summary of ThreadX (Azure RTOS) and Azure IoT C SDK

There was a different summary before the switch to the Eclipse Foundation. From the initial outset, competing with Amazon in the MCU space makes sense; but from a business perspective, how to accomplished this is a different story. It was a bit of a shock to see Microsoft make the acquisition, but the lack of direction and communication signaled problems. Windows CE had a whole ecosystem that grew up outside of Microsoft to provide support. Azure RTOS (ThreadX) has been around for a while, but the ecosystem isn't the same. After what Microsoft did to the Windows CE ecosystem, developers might not consider Azure RTOS as a dependable solution. In talking with a couple of clients about Azure RTOS, when I mention that Microsoft now has ThreadX (Azure RTOS), the reaction is more cautious. Their concern is what Microsoft is going to do with Azure RTOS. Now, we know. ThreadX requires a dedicated team that is very familiar with the product to maintain support. This dedication doesn't fit into Microsoft's matrix organizational structure. The business model of free licenses doesn't pay for engineers. There was no message of offering consulting support. In hindsight, ThreadX was not a fit for Microsoft, but at least the proper internal leadership took over and did the right thing to moved ThreadX to a community that can support it. The original developers of ThreadX started a new company and a new RTOS product, PX5 RTOS. They are offering consulting support so ThreadX will continue.

For the years that Microsoft controlled ThreadX, there was not much adult supervision on what was generated. The GSE is a complete mess to use as a starting point for design. Supporting different MCU manufacturers with one encompassing project that is built with Visual Code is not the best approach. The GSE was not the best marketing tool to promote Azure RTOS. The GSE brings up the question: "Was this project developed by a group of people who didn't talk to each other or one complete idiot?" There was a Visual Code craze for a period of time, so it explains why the GSE was developed. Visual Code is the wrong tool for MCU development when there are MCU-specific tools like STM32Cube. Microsoft eventually created examples for MCU toolchains, which was the right marketing and support direction, but they are not as complete as the GSE.

The smarter business and less costly approach to MCU support is the Azure IoT C SDK. The SDK can support different RTOSes and bare metal silicon. Like the GSE, there was a lack of adult supervision when it came to developing the Azure IoT C SDK. As demonstrated in chapters 9 and 10, it takes a great deal of effort to get the SDK integrated into the STM32 project. If you noticed that Chapter 12 didn't even cover the SDK for a wired connection, there was a struggle to get the DNS working. The SDK integration was dropped and the chapter was left to ping the network only. The Azure IoT C SDK intermixes the library with test tools and code. These items should be separate from one another with the test tools and code made as build options. The SDK should have been designed to easily drop into the MCU development tools, which explains why the SDK is not an option in STM32Cube software packages.

## 17.4 Final Summary

Even though it was a struggle at times, it was fun working on something other than Windows IoT Enterprise. Working with the STM32Cube tools and hardware was a refreshing experience. Creating the various projects exposed the little things about the tools and the intricacies of the MCU architecture, which I hope you found insightful. I do have some ideas for a second edition, so please let me know if you have any suggestions for future editions.

# A References

There are a number of references from websites to videos that were use to develop this book. The references are broken down into the chapters they pertain too.

Chapter 1

STM32 History: https://en.wikipedia.org/wiki/STM32

STMicroelectronics History: https://en.wikipedia.org/wiki/STMicroelectronics

STM32 Software Development Tools: https://www.st.com/en/development-tools/stm32-software-development-tools.html

Article: *From Express Logic ThreadX to Microsoft Azure RTOS* by Jonathan Blanchard, JBLopen Inc. October 14, 2020 https://www.jblopen.com/from-express-logic-threadx-to-microsoft-azure-rtos/

Chapter 2

STM32 development boards portfolio PDF:
https://www.st.com/resource/en/product_presentation/stm32_eval-tools_portfolio.pdf

Chapter 3

A bare metal programming guide: https://github.com/cpq/bare-metal-programming-guide

Article: *Bare-Metal STM32: From Power-Up To Hello World* by Maya Posch, Hackaday, August 21, 2024 https://hackaday.com/2020/11/17/bare-metal-stm32-from-power-up-to-hello-world/

Article: Bare-Metal STM32: *Exploring Memory-Mapped I/O And Linker Scripts* by Maya Posch, Hackaday, August 21, 2024 https://hackaday.com/2020/12/23/bare-metal-stm32-exploring-memory-mapped-i-o-and-linker-scripts/

Article: *STM32 EcoSystem (Development Environment) Setup* by Khaled Magdy, DeepBlue, https://deepbluembedded.com/stm32-ecosystem-development-environment-setup/

## Chapter 7

Video: *MOOC - Azure RTOS workshop - 6 ThreadX lab* by STMicroelectronics, 2022, https://www.youtube.com/watch?v=YQVoYlO8pbl

STMicroelectronics Knowledge Base article: *How can I add TraceX support in STM32CubeIDE?* By B.Montanari, September 30, 2021 https://community.st.com/t5/stm32-mcus/how-can-i-add-tracex-support-in-stm32cubeide/ta-p/49380

## Chapter 8

STMicroelectronics Community: *linker script help: What is meant by >RAM_D1 AT> FLASH*, April 2018, https://community.st.com/t5/stm32-mcus-products/linker-script-help-what-is-meant-by-gt-ram-d1-at-gt-flash/m-p/344897

Video: *STM32 MPU tips - 1 MPU usage in STM32 with ARM Cortex M7* by STMicroelectronics, 2020, https://www.youtube.com/watch?v=6IUfxSAFhlw&list=PLnMKNibPkDnEQXu4S6QUUHu SKj81MeqCz

STMicroelectronics Application Note: *Introduction to memory protection unit management on STM32 MCUs,* AN4838, 2024 https://www.st.com/resource/en/application_note/dm00272912-managing-memory-protection-unit-in-stm32-mcus-stmicroelectronics.pdf

## Chapter 11

Video: *MOOC - Azure RTOS workshop - 12 FileX integration into STM32Cube* by STMicroelectronics, 2022 https://www.youtube.com/watch?v=iKjS4G0bO_4

## Chapter 13

Video: *Dual Core Debugging on STM32H7 with STM32CubeIDE* by STMicroelectronics, 2021 https://www.youtube.com/watch?v=k3mXhPZSasw

STMicroelectronics Application Note: *STM32H745/755 and STM32H747/757 lines inter-processor communications*, AN5617 https://www.st.com/resource/en/application_note/an5617-stm32h745755-and-stm32h747757-lines-interprocessor-communications-stmicroelectronics.pdf

## Chapter 14

Video: *Inter-process Communication via OpenAMP* by STMicroelectronics, 2022 https://www.youtube.com/watch?v=MLcULDnF5ic

Video: *STM32 Dual Core #3. Inter core comm using OpenAMP and RPmsg || IPC || Shared Memory* by ControllersTech 2023
https://www.youtube.com/watch?v=hKiVYq0EaS4&list=PLflJKC1ud8gga7xeUUJ-bRUbeChfTOOBd&index=20

Video: *STM32 Dual Core #4. Inter core comm || FreeRTOS || OpenAMP || IPC || Shared Memory* by ControllersTech 2023 https://www.youtube.com/watch?v=miUzjoe3rAo&t=2s

Chapter 15

Website: *TouchGFX Tutorials Learn how to use TouchGFX and STM32Cube IDE* by ControllersTech https://controllerstech.com/touchgfx/

Training Course: *Accelerating HMI of Things with STM32 & TouchGFX MOOC* by STMicroelectronics https://www.st.com/content/st_com/en/support/learning/stm32-education/stm32-moocs/TouchGFX_webinar_MOOC.html

Video: *TouchGFX Webinar - 2 - Project on STM32H7B31-DK board* by STMicroelectronics, 2020
https://www.youtube.com/watch?v=XzwKXQ1RAtI&list=PLnMKNibPkDnHPh5mWtYkSLntmhvtZ4GyU&index=2

Video: *TouchGFX Presentation: How to start GUI development using TouchGFX* by STMicroelectronics, 2021 https://www.youtube.com/watch?v=OraJaTLn0CA

Video: *TouchGFX Presentation: How to Customize a TouchGFX Application Template for STM32H7B3I-DK: Part 1* by STMicroelectronics, 2021
https://www.youtube.com/watch?v=yiHjOH7zJP0

Video: *TouchGFX Presentation: How to Customize a TouchGFX Application Template for STM32H7B3I-DK: Part 2* by STMicroelectronics, 2021
https://www.youtube.com/watch?v=XnSg3UJfSFc

Video: *STM32 Graphics: How to use the slider widget to control a PWM output* by STMicroelectronics, 2021 https://www.youtube.com/watch?v=fzNEZIi2Gb0

Video: *Importing TouchGFX 4.16.0 project into STM32CubeIDE 1.6.1* by EE by Karl, 2021 https://www.youtube.com/watch?v=Q-mfuzW6S-0 - this one covers the renaming of aq project.

STMicroelectronics Community: *How to use TouchGFX?* By Iben.Thy, November 23, 2020 https://community.st.com/t5/stm32-mcus/how-to-use-touchgfx/ta-p/49553

Video: *TouchGFX Documentation* by STMicroelectronics, https://support.touchgfx.com/docs/introduction/welcome

STMicroelectronics Community: *Azure RTOS - Will there be full support (i.e. GUIX)?* December 14, 2020 https://community.st.com/t5/stm32-mcus-embedded-software/azure-rtos-will-there-be-full-support-i-e-guix/m-p/214477

Chapter 16

Website: *AZ3166 Mxchip IoT DevKit For prototype development of IoT and intelligent hardware* by MXCHIP® via ST Partner Program https://www.st.com/en/partner-products-and-services/az3166-mxchip-iot-devkit.html

MXCHIP Website: https://www.mxchip.com

Miscellaneous

Website: *IoT device development*, Microsoft https://learn.microsoft.com/en-us/azure/iot/iot-overview-device-development?WT.mc_id=IoT-MVP-5489

STMicroelectronics Community: *How to create a Thread using AzureRTOS and STM32CubeIDE* by B.Montanari September 16 2021 https://community.st.com/t5/stm32-mcus/how-to-create-a-thread-using-azurertos-and-stm32cubeide/ta-p/49418

STMicroelectronics Community Knowledge base: *How to redirect the printf function to a UART for debug messages* by ST AME Support NF, November 16, 2021 https://community.st.com/t5/stm32-mcus/how-to-redirect-the-printf-function-to-a-uart-for-debug-messages/ta-p/49865

Website: *Introduction to THREADX*, June 2022 https://wiki.st.com/stm32mcu/wiki/Introduction_to_THREADX

Video: Getting Started with X-CUBE-AZRTOS-H7 by STMicroelectronics, 2021 https://www.youtube.com/watch?v=OPVhsJfZA5A

# B Index

## A

**ABCOMTERM**, 18, 88, 112, 137, 202
**Azure IoT C SDK**, 3, 100, 115, 130, 316, 317
**Azure IoT Central**, i, 3, 4, 21, 76, 113, 115, 130, 135, 136, 137, 138, 144, 275
**Azure RTOS**, i, 2, 3, 4, 8, 20, 28, 31, 36, 51, 52, 54, 56, 75, 87, 93, 94, 101, 113, 115, 116, 144, 146, 150, 151, 160, 163, 172, 175, 198, 207, 209, 222, 231, 248, 264, 268, 275, 277, 285, 286, 300, 305, 315, 316, 317, 319, 320, 322

## B

**Binary**, 12, 40, 48, 276, 300, 301, 305, 313, 315
**Bootloader**, 41, 49

## C

**C/C++**, 81, 84, 99, 124, 193, 195, 224, 225, 246, 247, 260, 288, 295, 302, 306
**C++**, 233, 273
**char**, 220, 221, 222, 224, 243, 263, 291, 292, 293, 310, 311, 312
**Clock**, 1, 25, 28, 29, 32, 35, 36, 46, 54, 94, 117, 150, 163, 175, 178, 197, 203, 227, 285, 286, 298, 299, 313
**Compiler**, 81, 99, 124, 181, 182, 288
**CPU**, 26, 31, 41, 51, 176, 211, 270

## E

**Embedded Systems**, 1
**enum**, 103

## F

**FileX**, 3, 4, 8, 92, 145, 146, 147, 148, 149, 151, 152, 153, 155, 157, 158, 160, 162, 164, 165, 166, 167, 169, 170, 231, 316, 320
**Firmware**, 1, 15, 41, 49, 75, 79, 80, 85, 87, 89, 90, 115, 135, 136
**float**, 83, 84, 129, 291, 292, 295, 310, 311, 312

## G

**GPIO**, 32, 33, 36, 47, 59, 83, 106, 129, 188, 192, 193, 197, 202, 205, 212, 229, 278, 281, 282, 283, 284, 294, 296, 307, 308
**Graphics**, 321
**GUIX**, 3, 8, 20, 23, 231, 232, 233, 248, 249, 250, 251, 252, 253, 257, 258, 259, 260, 261, 262, 263, 264, 265, 266, 268, 269, 270, 271, 273, 315, 322

## H

**Heap**, 182
**HTS221**, 76, 77, 79, 83

## I

**I/O**, 1, 11, 25, 26, 32, 34, 35, 36, 43, 45, 46, 52, 76, 83, 93, 101, 116, 145, 146, 159, 172, 185, 186, 187, 210, 213, 229, 275, 276, 278, 319
**I2C**, 7, 26, 77, 85, 278, 279, 289, 290, 293, 313

## L

**LCD**, 7, 8, 9, 229, 231, 233, 244, 246, 272, 273
**LED**, 31, 33, 34, 36, 37, 41, 43, 46, 47, 49, 51, 52, 55, 57, 58, 59, 60, 61, 63, 64, 65, 69, 72, 121, 122, 144, 185, 188, 191, 193, 198, 228, 229,

277, 280, 281, 283, 289, 294, 295, 296, 297, 307, 308, 309
**LPS22HB**, 76, 77

# M

**MEMS**, 76, 77, 79, 80
**Microprocessor**, 18
**Multimedia**, 188, 215
**MXCHIP**, 8, 9, 275, 276, 286, 287, 288, 291, 298, 300, 301, 302, 303, 305, 306, 315, 322

# N

**NetX Duo**, 3, 87, 89, 90, 91, 92, 97, 101, 107, 109, 111, 113, 115, 116, 123, 124, 130, 144, 151, 171, 172, 173, 174, 179, 181, 183, 184, 231, 277, 300, 306, 316

# O

**OpenAMP**, 209, 213, 218, 219, 221, 222, 223, 228, 229, 320, 321
**OpenOCD**, 277, 296
**Operating System**, i, 2
**Overrides**, 102, 266, 267, 272

# P

**Package**, 13, 16, 56, 65, 76, 89, 91, 285, 300
**Pixel**, 248
**Pointers**, 58, 62, 154, 168, 203, 263, 265
**Preprocessor**, 181, 182
**Protected**, 244
**PWM**, 280, 281, 282, 283, 321

# R

**RAM**, 39, 67, 145, 146, 149, 152, 154, 157, 165, 170, 182, 183, 203, 204, 269, 320
**Recursive**, 22

# S

**Scope**, 25, 135, 137, 176, 184, 231
**SDK**, 89, 98, 100, 124, 316, 317
**Serial**, 8, 18, 26, 61, 63, 75, 83, 84, 88, 112, 137, 156, 170, 183, 197, 202
**SPI**, 7, 26, 88, 103, 106, 112, 137, 171, 181, 276

**Stack**, 3, 58, 59, 62, 68, 87, 90, 107, 108, 109, 122, 123, 162, 170, 179, 182, 262, 263, 264, 295, 296, 309, 310
**ST-Link**, 7, 8, 23, 145, 219, 277, 296
**STM32CubeIDE**, i, 2, 11, 12, 13, 14, 15, 19, 22, 34, 41, 43, 44, 49, 55, 56, 57, 65, 76, 80, 82, 85, 87, 88, 92, 96, 97, 98, 101, 110, 112, 115, 119, 120, 124, 126, 129, 130, 133, 137, 145, 151, 152, 164, 165, 166, 179, 189, 198, 217, 218, 219, 232, 235, 241, 244, 268,271, 273, 286, 287, 289, 291, 298, 300, 301, 303, 306, 315, 320, 321, 322
**STM32CubeMX**, 1, 11, 12, 13, 15, 16, 31, 34, 35, 41, 44, 46, 51, 56, 75, 76, 77, 89, 92, 96, 97, 116, 119, 129, 130, 134, 145, 151, 152, 158, 159, 165, 170, 171, 179, 185, 189, 190, 209, 217, 232, 235, 241, 242, 272, 273, 277, 287, 289, 303, 315
**String**, 59, 103, 153, 167, 206, 209, 220, 222, 228, 243, 291, 309
**Strings**, 253
**struct**, 220, 221, 222, 223, 224, 266
**switch-case**, 265

# T

**Thread**, 19, 51, 58, 59, 61, 62, 63, 64, 65, 66, 67, 68, 72, 73, 107, 108, 113, 119, 122, 123, 151, 152, 153, 154, 155, 157, 162, 166, 167, 169, 170, 179, 180, 200, 201, 202, 205, 206, 209, 216, 217, 220, 221, 222, 223, 228, 229, 262, 263, 264, 265, 271, 289, 295, 296, 309, 310, 313, 316, 322
**ThreadX**, i, ii, 1, 2, 3, 4, 5, 8, 20, 22, 25, 31, 36, 49, 51, 52, 54, 56, 57, 58, 59, 62, 64, 65, 68, 72, 76, 85, 87, 89, 92, 93, 94, 107, 108, 113, 116, 117, 121, 122, 123, 132, 145, 146, 147, 151, 160, 170, 171, 172, 173, 198, 200, 201, 207, 209, 216, 217, 218,231, 232, 233, 248, 251, 258, 263, 264, 265, 268, 273, 275, 277, 285, 296, 309, 310, 315, 316, 317, 319, 320
**Timer**, 28, 188, 198, 213, 280, 286
**TouchGFX**, 19, 20, 231, 232, 233, 234, 235, 236, 239, 241, 242, 243, 244, 273, 315, 321
**TraceX**, 3, 19, 65, 66, 67, 68, 70, 71, 72, 73, 231, 320
**typedef**, 103, 266

## U

**UART**, 2, 26, 36, 46, 57, 59, 60, 64, 84, 110, 127, 152, 153, 166, 179, 180, 191, 192, 193, 219, 280, 290, 292, 293, 295, 322
**USBX**, 3, 23, 231, 315

## V

**Visual Studio**, 3, 20, 23, 115, 248, 250, 253, 257, 258, 260, 262

## W

**web**, 20, 22, 91, 100, 125, 249, 250, 315
**while-loop**, 36, 46, 83, 191, 193, 202
**Workspace**, 12, 20, 34, 44, 56, 76, 96, 100, 119, 125, 151, 165, 179, 189, 287